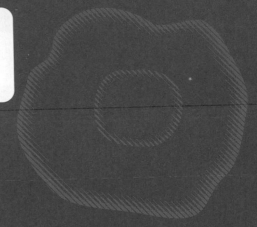

Blue light wavelength: 1.9×10^{-5} inch

Interstellar dust grain: diameter 4×10^{-6} inch

Cell: diameter 5×10^{-4} inch

Black hole: diameter 40 miles

Large moon crater: diameter 120 miles

Largest asteroid: diameter 620 miles

Mars: diameter 4,223 miles

White dwarf: diameter 5,000 miles

Venus: diameter 7,521 miles

WORKINGS
OF THE UNIVERSE

TIME
LIFE
BOOKS

Other Publications:
LOST CIVILIZATIONS
THE NEW FACE OF WAR
HOW THINGS WORK
WINGS OF WAR
CREATIVE EVERYDAY COOKING
COLLECTOR'S LIBRARY OF THE UNKNOWN
CLASSICS OF WORLD WAR II
TIME-LIFE LIBRARY OF CURIOUS AND UNUSUAL FACTS
AMERICAN COUNTRY
THE THIRD REICH
THE TIME-LIFE GARDENER'S GUIDE
MYSTERIES OF THE UNKNOWN
TIME FRAME
FIX IT YOURSELF
FITNESS, HEALTH & NUTRITION
SUCCESSFUL PARENTING
HEALTHY HOME COOKING
UNDERSTANDING COMPUTERS
LIBRARY OF NATIONS
THE ENCHANTED WORLD
THE KODAK LIBRARY OF CREATIVE PHOTOGRAPHY
GREAT MEALS IN MINUTES
THE CIVIL WAR
PLANET EARTH
COLLECTOR'S LIBRARY OF THE CIVIL WAR
THE EPIC OF FLIGHT
THE GOOD COOK
WORLD WAR II
HOME REPAIR AND IMPROVEMENT
THE OLD WEST

This volume is one of a series that
examines the universe in all its aspects,
from its beginnings in the Big Bang to the
promise of space exploration.

Evidence for the complex structure of
the universe comes in a false-color photo-
graph of particle tracks in a bubble
chamber. The collision of a negative
kaon *(yellow)* and an invisible proton
yields a positive pion *(orange),* a negative
pion *(purple),* and an invisible lambda
particle, one member of the family of particles that includes familiar protons and
neutrons. The presence of the neutral
lambda is revealed only by its V-shaped
decay into a proton *(red)* and a muon
(blue), which immediately decays into an
electron *(large green spiral)* and two invisible neutrinos.

VOYAGE THROUGH THE UNIVERSE

WORKINGS OF THE UNIVERSE

BY THE EDITORS OF TIME-LIFE BOOKS
ALEXANDRIA, VIRGINIA

CONTENTS

1 6 LESSONS IN UNITY
38 THE ORIGIN OF MATTER

2 50 IN SEARCH OF QUANTUM GRAVITY
79 THE SUPERSTRING UNIVERSE

3 90 ON THE MEANING OF NOW
125 THE RIDDLE OF TIME'S PASSAGE

134 GLOSSARY

136 BIBLIOGRAPHY

139 INDEX

143 ACKNOWLEDGMENTS

143 PICTURE CREDITS

The lights of a moving car trace a circular course 1.3 miles in diameter on the Illinois prairie—marking the vast underground Tevatron accelerator at the Fermi National Accelerator Laboratory near Chicago. In their quest to unify the four fundamental forces of nature, physicists use devices like the Tevatron accelerator to smash particles together in violent collisions simulating the high-energy conditions that existed in the earliest instants after the Big Bang.

ne of the greatest discoveries in the history of science seems, by all accounts, to have had its origin in the most ordinary of events. The year was 1666, and for some time England had been suffering through a particularly nasty episode of the bubonic plague. Because the university at Cambridge had been closed to prevent the disease from spreading throughout the student population, a young scholar named Isaac Newton, who had received his bachelor's degree in April of 1665 and was about to begin postgraduate work, had been forced to continue his studies at home, in the village of Woolsthorpe, about sixty miles northwest of Cambridge. Lodged at his desk overlooking an apple orchard on the family farm, he immersed himself in his favorite subject, mathematics, employing a new form of it—now called calculus—that he himself had invented to investigate the behavior of objects in motion, especially heavenly bodies such as the Moon and the planets.

Some fifty years earlier, German astronomer Johannes Kepler had revolutionized the study of celestial mechanics with the first accurate descriptions of planetary motions, but as yet no one had grasped the underlying principle that explained why the planets behave as they do—why, for example, those farther from the Sun move more slowly in their orbits than do planets closer in. One day, probably late in the summer of 1666, Newton had a sudden and profound insight that solved the mystery. The exact circumstances of this flash of comprehension are lost to history, but many years later, the mathematician supposedly recalled that he had been inspired by witnessing an apple falling from a tree outside his window.

What Newton realized was that the force of gravitation that causes an apple to drop to the ground also holds the Moon in thrall, maintaining it in its orbit. An object hurled parallel to Earth's surface, he argued, is subject both to the force that propelled it and to the pull of gravity, and it is therefore continuously moving both horizontally and downward, tracing a curved path. It eventually strikes the ground because gravity is stronger than the propulsive force, causing the object to fall faster than it travels horizontally. If gravity were weaker than the force of propulsion, the object would follow a straighter path and would be carried out into space as the planet's surface curved away from it. But—and here was the crucial insight—if the two forces were precisely balanced, the distance the object fell would be offset by its horizontal progress, and it would never reach the ground: It would be in orbit around Earth, like the Moon.

Newton proceeded to test his supposition mathematically. He reasoned that gravitational force was commensurate with mass and that the gravity of large bodies such as Earth could be thought of as emanating from a point at their center. Assuming, as no one had before, that gravity extends out into space, he deduced that its strength ought to diminish in proportion to the square of the distance over which it propagates, a principle now known as the inverse-square law. According to estimates of the day, the Moon was 240,000 miles from the center of the Earth, about 60 times farther than was an apple on a tree; it was therefore subject to 3,600 times less gravity. Knowing how strong gravity was at Earth's surface—accelerating an object by 32 feet per second for every second it was falling, as Galileo had determined—he calculated how quickly the Moon would have to circle Earth in order to offset gravity's weaker pull at that distance and stay in orbit. As it happened, Newton's result, 29.3 days, was discouragingly off the observed time of 27.3 days.

Several years later, however, scientists arrived at a new, more accurate estimate of Earth's radius, revising it upward. The revision brought Newton's calculations into line, thereby confirming that gravitation was not just a terrestrial but a universal phenomenon. It accounted for the Moon's orbit as well as for the orbits of all the planets around the Sun. Kepler's findings now made complete sense: The outer planets move more slowly because the Sun's gravity is weaker there.

Posing without the wig usually prescribed by conventions of the day, Isaac Newton gazes into the middle distance in a portrait painted in 1689, two years after he published his *Principia,* the landmark book describing his theory of universal gravitation.

Newton's theory of universal gravitation, as he would formally designate it in *The Mathematical Principles of Natural Philosophy,* or *Principia,* published in 1687, did more to explain the basic workings of the universe than any of the ideas that preceded it, and its enduring influence on the course of scientific inquiry cannot be overstated. It stood unchallenged for more than 200 years and laid the foundation for the great twentieth-century theories that would eventually supplant it. His formulas still hold true for describing the majority of physical events throughout the cosmos; only at extremes of velocity and mass do the equations of Albert Einstein's theories of relativity offer a more accurate picture.

The theory of universal gravitation—a work of unparalleled significance—followed in a tradition of long standing. Science has always searched for the fundamental principles underlying complex and diverse physical phenomena, trying, in effect, to reduce nature to its lowest common denominators. Such attempts to unify disparate realities trace all the way back to the pronouncements of Aristotle that the entire world was composed of just four elements: earth, air, fire, and water. Part of Newton's genius stems from this urge toward unification, which led him to see that falling apples and orbiting planets are simply different manifestations of the same physical law.

In the twentieth century, those who search for unity have met with both great success and great frustration. Early on, physicists posited that all matter consists of a trio of irreducible elementary particles—the proton, the neutron, and the electron—which come together in intricate, elegant arrangements to fashion the whole panoply of atoms and molecules that make up the material world. Subsequent experiments and theoretical elaborations have complicated the picture, identifying additional fundamental constituents and a host of other entities that come into being only under special circumstances. However, particles are less the focus of unification efforts than are the forces that dictate how these bits of matter interact. By midcentury, physicists had discovered that only four basic forces are responsible for all such interactions: gravity, electromagnetism, the strong nuclear force (which binds protons and neutrons together in the atomic nucleus), and the weak nuclear force (which controls radioactive decay). In an effort to discover an even more fundamental description of how the universe operates, researchers have managed to find the necessary connections between at least two, and perhaps three, of the forces. But gravity—the first to be named—has turned out to be a stumbling block to complete unification.

Since the 1920s, physics has relied on two distinct languages to describe the workings of the cosmos: relativity and quantum mechanics. Simply put, relativity looks at the universe on the large scale, examining among other things the role of gravity in shaping the overall structure of space. Quantum mechanics, for its part, concerns the microcosmic world of particles. Although at first narrowly applied to the behavior of the electron, quantum theory soon achieved a broader scope and eventually led to the concept that all of the forces might be described in terms of particles, making the possibility of a grand synthesis all the more believable. Researchers pursuing the idea have succeeded in constructing mathematical models for the quantum nature of three of the forces, models that have been confirmed by laboratory experiments and observations.

However, gravity still refuses to cooperate, in part because it is vanishingly weak at the small-scale level of quantum mechanics. A thoroughly convincing particle theory of gravitation has yet to emerge, and it seems likely that theorists hoping to fit all of nature into a single nutshell will have to revamp the basic vocabulary of physics.

THE ELECTRIC-MAGNETIC CONNECTION

For about 150 years after Isaac Newton's great elucidation in *Principia,* gravity was the only fundamental force to have been worked out in precise equations and formulas. The next force to be explained—even more clearly a product of the search for unification—linked three well-known, if not entirely well-understood, phenomena: electricity, magnetism, and light itself.

Long before intricate mathematical theories were devised to describe electric and magnetic forces, observers had noted their effects in the world, routinely associating them with specific materials. The ancient Greeks, for example, found that rubbing amber against fur was a particularly good way to generate what is now known as static electricity. (The word "electricity" comes from elektron, the Greek word for amber.) An awareness of magnetism may be even older, probably dating to the Iron Age and the discovery that a certain type of rock—lodestone, or magnetite in modern parlance—would attract fragments of iron. Over time, intriguing similarities between the two powers emerged. Early on, the observant noticed that any magnetized object had two ends, or poles (so named because a suspended magnet would always align with one end pointing north and the other south), and that like poles repelled each other and opposite ones attracted. Trial-and-error experiments with static electricity revealed an equivalent characteristic: With friction, objects could be endowed with an electric charge that was either positive or negative, such that objects with opposite charge were pulled together and those with the same charge pushed apart.

Toward the end of the eighteenth century, a further congruity became evident. In 1785, French physicist Charles Augustin Coulomb demonstrated that both electricity and magnetism obeyed Newton's inverse-square law, losing strength in proportion to the square of the distance. Investigators now began to focus on how these forces exerted their influence through space and on how they might interact, given that they were so much alike.

In the early part of the nineteenth century, Danish scientist Hans Christian Ørsted took a significant step toward the unification of electricity and mag-

The efforts of Michael Faraday (near right) and James Clerk Maxwell produced the unification of electricity, magnetism, and light. In a pathbreaking work published in 1839, Faraday demonstrated the reciprocal relationship between electricity and magnetism and proposed that both were conveyed through space by invisible lines of force. Inspired by Faraday's ideas, Maxwell constructed, in 1873, a comprehensive mathematical theory of electromagnetism, proving that electricity and magnetism were two aspects of the same force. Then, in a brilliant deduction, he concluded that light was a form of electromagnetic radiation.

netism, almost in spite of himself. In 1807, suspecting that the shared attributes of the two forces were more than just coincidence, he devised an experiment to see if a moving electric charge—that is, an electric current—might somehow affect a magnet, although a static charge clearly did not. He placed a wire across a compass needle, perpendicular to the axis of its poles, and sent current flowing through the wire, hoping the needle would swing round as it would have if a bar magnet had been so positioned. The needle did not move, and Ørsted assumed that he had been wrong. However, he continued to investigate the properties of electricity and magnetism in his role as a lecturer at the University of Copenhagen. In 1820, thirteen years after his initial experiment, while teaching a class about electric currents, he accidentally placed a current-bearing wire parallel to a compass needle. Much to his amazement, the needle immediately shifted so that it pointed perpendicular to the wire. The current was indeed creating a region of magnetism around it; Ørsted had only been mistaken, and quite stubbornly so, about the direction in which the force was operating. But neither he nor anyone else at the time could explain the odd alignment.

A further correspondence, which would not emerge for more than ten years, owed to the efforts of a somewhat more imaginative experimenter. Englishman Michael Faraday was a blacksmith's son and had grown up in poverty, with only a bare minimum of education. At the age of twelve, however, he was apprenticed to a bookbinder, which turned out to provide fertile ground for his innate intellectual curiosity. In his spare time, Faraday would pore over texts that had come in to be bound and was particularly fascinated by works on science. Years later, at the prompting of a friend, he attended a series of lectures by the prominent scientist Humphry Davy of the Royal Institution, one of the leading scientific establishments of the day. Faraday was so inspired by the experience that he carefully copied his notes from the talks, bound them in leather, and sent them to Davy as a gift—and, in effect, as an application for a job as a laboratory assistant. Davy was hesitant at first but eventually offered Faraday a position.

The enthusiastic young man would prove to be one of the most dedicated scientific investigators of his time, working at the Royal Institution for forty-six years and becoming its director after Davy. His first major accomplishment, achieved in 1831, was to show that, just as an electric current produces magnetism, a source of magnetism set in motion creates electricity. He constructed a simple apparatus consisting of a magnet that could be moved through a coil of wire attached to registers. Whenever the magnet was shifted, a weak current would flow through the wire, and it would stop whenever the magnet was stopped. Electricity and magnetism now seemed to be more inseparable than ever.

Faraday was spurred to learn more about how the two forces communicated their effects through space. Many scientists had opted for the somewhat unsatisfactory explanation that they operated through a kind of mystical action-at-a-distance, without any intervening medium. Faraday's observa-

A GRAVITATIONAL MYSTERY SOLVED

In the late seventeenth century, Isaac Newton defined the basic principles that govern the orbits of bodies in space *(below)* when he set forth his laws of motion and his universal law of gravitation. Newton established, among other things, that two bodies attract each other with a force that is proportional to the product of their masses, and that the attraction varies inversely with the square of the distance between them. This "inverse-square law" says two bodies four million miles apart feel one-quarter the pull experienced by bodies of the same mass only two million miles apart.

As astronomers after Newton scrutinized the orbital motions of the planets and moons in the Solar System, they found that any discrepancies in the movements eventually turned out to be the result of the influence of previously unknown bodies. Such anomalies in the orbit of Uranus, for example, led to the discovery of Neptune in 1846. Mercury, however, proved to be a bothersome exception. The perihelion of Mercury—its closest approach to the Sun—was seen to precess, or move forward slightly on each orbit *(bottom)*, by an amount that could not be accounted for by Newton's laws alone. The mystery remained unsolved until 1915, when Albert Einstein's revolutionary concept of gravity redefined space itself *(pages 14-15)*.

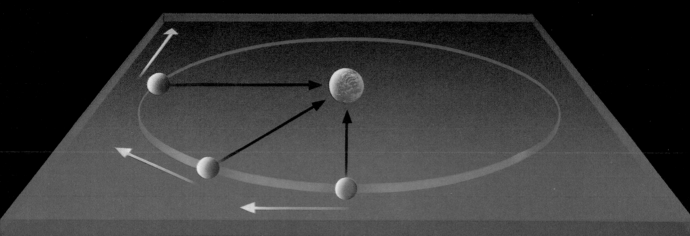

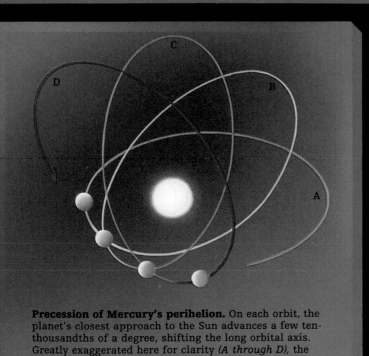

Newton demonstrated that the orbit of one body about another—in this case, the Moon about Earth—is the result of two forces, or vectors, acting at right angles. If no other force were present, the Moon's orbital momentum *(white arrows)* would carry it off on a tangent into space. But gravity *(black arrows)* counters that tendency by drawing the Moon toward Earth. The result is a compromise: a curving path that balances the forces of momentum and gravity and holds the Moon in orbit.

Precession of Mercury's perihelion. On each orbit, the planet's closest approach to the Sun advances a few ten-thousandths of a degree, shifting the long orbital axis. Greatly exaggerated here for clarity *(A through D)*, the small precessions are greater than Newton's laws allow.

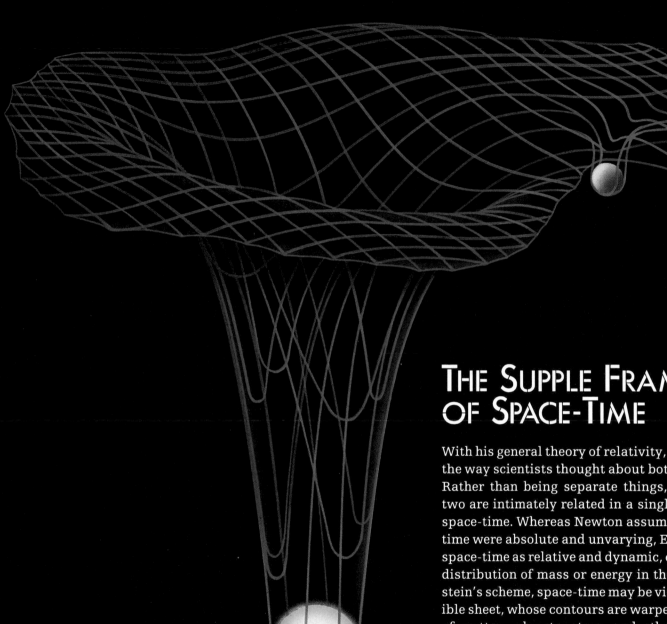

The Supple Framework of Space-Time

With his general theory of relativity, Einstein changed the way scientists thought about both space and time. Rather than being separate things, he declared, the two are intimately related in a single entity he called space-time. Whereas Newton assumed that space and time were absolute and unvarying, Einstein described space-time as relative and dynamic, changing with the distribution of mass or energy in the vicinity. In Einstein's scheme, space-time may be visualized as a flexible sheet, whose contours are warped by the presence of matter—planets, stars, and other bodies. (In the illustration above, a grid of imaginary lines marks coordinates that help reveal the warping.)

In a bold break with Newtonian theory—which regarded gravity as an independent force—Einstein declared that the observed effects of gravity are merely the consequence of the distortions produced by matter. The deformations, called gravity wells, vary with the mass of the object. For example, Mercury's gravity well *(above)* is a mere dimple compared to the Sun's *(left)*, which is about 10 million times greater.

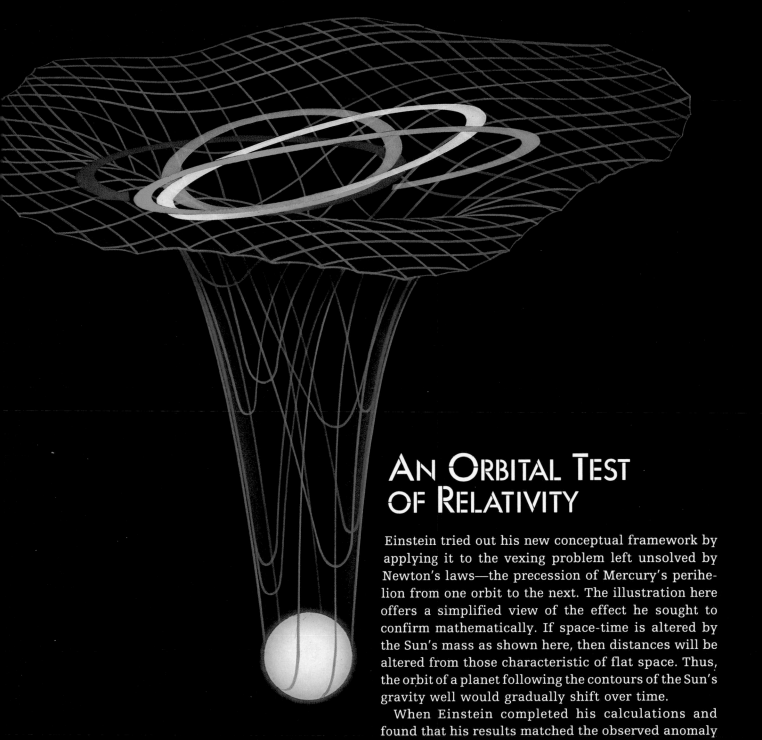

AN ORBITAL TEST OF RELATIVITY

Einstein tried out his new conceptual framework by applying it to the vexing problem left unsolved by Newton's laws—the precession of Mercury's perihelion from one orbit to the next. The illustration here offers a simplified view of the effect he sought to confirm mathematically. If space-time is altered by the Sun's mass as shown here, then distances will be altered from those characteristic of flat space. Thus, the orbit of a planet following the contours of the Sun's gravity well would gradually shift over time.

When Einstein completed his calculations and found that his results matched the observed anomaly in the orbit of Mercury, he felt confident that he had transcended the bounds of Newtonian physics. "For a few days," he later wrote, "I was beside myself with joyous excitement."

UNCERTAIN DESCENT INTO A GRAVITY WELL

Einstein's concept of gravity explains observations conducted at the large scales of everyday experience. But gravity is such a weak force that at the subatomic scales of quantum mechanics, relativity simply does not apply. Quantum theory contains an element of uncertainty because it involves elusive particles whose activity defies precise measurement. When a gravity well is viewed at quantum scales—on the order of 10^{-33} centimeter, say, or 10^{-43} second—the grid lines defining the structure of the space-time sheet become less distinct, with coordinates fluctuating around their average values, so that space-time itself becomes subject to uncertainty. In quantum physics, then, subatomic particles trace paths that may be expressed only as sets of probabilities, rather than in the definite terms of Einsteinian relativity.

Quantum theorists have also gone beyond Einstein in postulating a particle that carries gravity, the graviton (below). Directly confirming its existence will be difficult, however, because the graviton is thought to operate with very faint effect over distances as much smaller than the nucleus of an atom as the atom is smaller than the Sun.

The illustrations at right represent a theoretical view—greatly exaggerated for clarity—of subatomic activity within the Sun's gravity well. Near the top of the well, far from the mass of the Sun, space-time is relatively flat and gravitons (wavy lines) are relatively scarce. Closer to the Sun's mass, the structure of space-time at the quantum level would be more uncertain—represented here by fuzzier grid lines—and there would be increasing interactions between gravitons and matter-carrying particles such as quarks (pie slices) and electrons (balls).

tions had led him to believe in something more tangible if no less astounding. In a familiar experiment now widely performed in schoolrooms and first described by René Descartes in 1644, Faraday sprinkled iron filings on a piece of paper lying over a magnet and studied the patterns that resulted. After some 15,000 similar and more complex tests with electric charges and currents as well as magnets, he concluded that both electricity and magnetism were conveyed by invisible but real lines of force that flowed out into space, interacting with force lines from other sources to produce magnetic and electric effects. He described the lines as forming and mapping out a so-called field for any given force. The concept was revolutionary. It provided a whole new way of thinking about these forces, in terms of the extended regions over which they had influence.

MATHEMATICIAN TO THE RESCUE

Faraday first published his results in 1839 in a paper titled "Experimental Researches in Electricity," a work that he continued to revise over the next fifteen years as he garnered more information. Because of his poor education, however, he lacked the mathematical skills to put his findings into equations that would spell out the fundamental nature of electricity and magnetism, as well as their inherent relatedness.

Into the breach stepped James Clerk Maxwell, a Scottish math prodigy who had won a mathematics award at Edinburgh Academy when he was fourteen. Studying at the University of Cambridge in the early 1850s, he read one of the later versions of Faraday's paper and became intrigued by the notion of fields and force lines. Maxwell realized that there were certain similarities between what Faraday had described and the mechanics of liquids. He followed up on this clue, constructing formulas that described the lines according to the principles of hydrodynamics, or fluid flow. The lines could be thought of as tubes carrying an electric fluid whose velocity corresponded to amperage and whose pressure represented electric potential, or voltage. The mathematically calculated action of this fluid worked well to explain known electrical effects, and Maxwell's 1856 paper on the subject was greeted with admiration, not least from Faraday himself, who wrote to Maxwell, "I was at first almost frightened when I saw such mathematical force made to bear upon the subject, and then wondered to see that the subject stood it so well."

Maxwell's supreme contribution was yet to come. Working diligently both at the blackboard and in the laboratory, he went about developing a mathematical description that finally united the phenomena of electricity and magnetism into one, through Faraday's field concept. At the heart of the description were four equations, the two most significant of which detailed the production of a magnetic field by an electric current, and of a current and its accompanying electric field by the changing magnetic field that results when a magnet is moved. The two equations ought to have been mathematically consistent, but according to Maxwell's initial calculations they were not. He eventually discovered that the formula describing how a magnetic field arises from an electric current was missing a key term that would make the equations fully compatible. The term related to a physical effect that had not previously been considered: producing a magnetic field not just by generating a steady current but by varying a current in some way—reversing its direction, say, or strengthening or weakening it—and thereby changing its associated electric field. Experiments confirm Maxwell's inclusion of the additional term: A compass needle placed between two charged plates swings around as the electric field between the plates is altered by modifying the current to the plates.

The insight was a crucial one. To begin with, it demonstrated that electric and magnetic fields were so intricately associated—each acting as a source for the other—that it made more sense to talk of a single electromagnetic phenomenon. In his 1873 paper entitled "Treatise on Electricity and Magnetism," Maxwell showed that such an electromagnetic field was made up of an electric field in one plane and a magnetic field in a perpendicular plane (finally explaining Hans Ørsted's experience). Solutions to his four equations now clearly proved that electricity and magnetism were just different aspects of the same basic force.

A further implication followed. Since an electric field could generate a magnetic field and vice versa, they could sustain each other and propagate through space far beyond their originating charge or magnetic source. For example, an electric field that began to change would give rise to a magnetic field, which would vary in response to the electric field's changes and thereby generate another electric field, and so on. Maxwell realized that a typical form of this propagation would be a wave, and when he used his equations to calculate its speed, he came up with a familiar figure, the velocity of light. Boldly—and correctly—he concluded that light was a form of electromagnetic wave and thus also fit into the picture. In addition, he reasoned that electromagnetic waves were bound to exist at all sorts of wavelengths, both shorter and longer than those of light. Scientists now know of the full spectrum of electromagnetic energy, from the microscopically short wavelengths of gamma rays to the yards-long measures of radio waves, first produced by Heinrich Hertz in 1889, ten years after Maxwell's death. By that time, electromagnetism had without quarrel earned the right to be known as a fundamental force of nature.

A NEW PERSPECTIVE

Over the next fifty years, physics was to become an increasingly complex science, as new and frequently bizarre concepts transformed long-held notions about how the world worked. In 1897, for example, the electron made its entry when English scientist Sir Joseph John Thomson, director of the prestigious Cavendish Laboratory, succeeded in demonstrating that an electric current was actually a stream of negatively charged particles. A few years later, German physicist Max Planck set the stage for quantum mechanics with his discovery that accelerated electrons emit radiation in discrete amounts, or quanta, rather than across a continuous range of values. In 1905, Einstein elaborated on Planck's work, proposing that light consists of particles—later dubbed photons—rather than waves. (As perplexing as it seems, physicists now acknowledge that both the wave and the particle descriptions of light are valid.) Ultimately, these revelations would profoundly alter the tenets of electromagnetism. More immediately, however, Maxwell's elegant theory played its own part in the overturning of traditional assumptions. One of its offspring was relativity.

Also introduced by Einstein in 1905 and immediately recognized as the cornerstone of a whole new perspective, the special theory of relativity arose from the need to resolve a puzzling discrepancy between classical Newtonian mechanics and the world of electromagnetic fields and waves. Maxwell's equations indicated that the velocity of electromagnetic waves through space is absolute, unaffected by any motion of the object generating them. According to classical physics, this was nonsensical: A beam of light from the headlamp of a speeding train, for example, ought to travel at a rate that equals the speed of light plus the train's velocity; such a beam would travel faster than a light beam from a stationary lamp. Through a series of ingenious "thought experiments" that were eventually confirmed by observations, Einstein managed to prove Maxwell's view correct; the speed of the light beam is constant, regardless of the motion of its source or of any observer.

The consequences of this proposition are what make relativity so hard to grasp. To begin with, the speed of light, like any velocity, is a measure of distance traveled over time. If the light beam from a speeding train travels no faster than a beam from a stationary train, it must be because the scales of time and distance for each train are different and are relative to each train's motion; thus, according to relativity, there is no absolute standard for time or distance that applies everywhere in the universe. The special theory—so called because it involves the special case of uniform, or nonaccelerated, motion—also indicates that an object's mass is directly proportional to its kinetic energy, or the amount of energy represented by its motion. At the speed of light, an object would be infinitely massive and so unmovable. Therefore, no mass could be propelled that fast, let alone any faster. This relationship between mass and energy led Einstein to the profound observation that the two are equivalent; one can change into the other under the right conditions, according to the dictates of the century's most famous equation, $E = mc^2$, or energy equals mass times the speed of light squared.

Although Maxwell's theory had set the stage for Einstein's seminal pronouncements, it was beginning to outgrow its original form as a result of investigations at the opposite end of the scale—the submicroscopic world of the atom. Not long after his discovery of the electron, Cavendish Laboratory's Thomson had developed a rudimentary theory of matter that depicted the atom as a positively charged cloud seeded with negatively charged particles, like raisins in a plum pudding. It would take more than a decade for two great physicists—New Zealander Ernest Rutherford and Danish theorist Niels

Bohr, both of whom spent time at Cavendish—to arrive at more accurate models. First, Rutherford postulated that atoms actually consist of a dense, positively charged nucleus orbited by a whirling array of electrons. Bohr refined the picture in 1913 by incorporating Planck's quantum principle: Electrons were restricted to specific orbits defined by their energy level, and they moved to higher or lower orbits by absorbing or emitting a discrete amount of electromagnetic energy in the form of photons.

According to this view, atoms were held together by the attraction of positive and negative charges, and all molecular structures and chemical reactions could be explained in terms of the interaction of charged particles. The electromagnetic force was thus clearly central to the very constitution of matter and intimately involved in all of its manifestations. But field theory as it was formulated by Maxwell was not quite up to the challenge of predicting all the ramifications of this new conception. Quantum mechanics was on the verge of making abundantly clear just how much adjusting the old equations needed.

THE QUANTUM WAY

Bohr's picture of the atom has survived in the popular imagination to this day, but his contemporaries were quick to recognize that it did not by itself do justice to the intricacies and oddities of what must be going on at the subatomic level. The electron in particular was turning out to be a curious study. Most intriguing was its ability to move from one orbit to another by a so-called quantum leap, making the jump without traversing the space in between. Impossible as this seems, it was, of course, the central point of quantum physics, that energy shifts take place in discrete steps. By the early 1920s, investigators were finding that other characteristics of the electron besides its energy are also quantized—that is, occur only at specific values, identified by so-called quantum numbers—including its charge, its angular momentum, and its spin rate around its own axis. The goal became to formulate a mathematical model whose equations would describe all the mechanics of an electron's quantum behavior, no matter how bewildering.

A plausible theory of quantum mechanics finally emerged in 1927, but only after several earlier attempts had fallen short of the mark. The central figure in the investigations was Werner Heisenberg, a young German physics teaching fellow at the University of Göttingen. Heisenberg had begun pursuing the subject early in the 1920s because he found Bohr's concept of orbiting electrons to be imprecise at best and misleading at worst; it was wrong, he felt, to think of electrons as tiny planets, because planets clearly could not jump from one orbital radius to another. He opted for a less vivid description, in which electrons occupy quantum states that represent distinct energy levels, transitioning between these states—here he did not differ from Bohr— through the emission or absorption of photons. In 1925, along with his faculty adviser, Max Born, and fellow student Pascual Jordan, he developed a complex mathematical technique that predicted values for energy emissions from atoms in transition, values that matched those actually obtained through laboratory experiments. However, the equations seemed to break one of the basic rules of mathematics, known as the commutative law, which states that no matter the order in which two numbers are multiplied, the result is always the same. In Heisenberg's equations, multiplying a value for an electron's position and a value for its momentum produced different results depending on which value was calculated first. Heisenberg made little of the anomaly in papers on the subject, but something definitely had to be clarified before this version of quantum mechanics could stand.

The clarification resulted from what might at first seem to be an additional complicating factor. In 1924, an obscure French theoretician, Louis de Broglie, had proposed that physicists might find it more useful to think of electrons as waves than as particles. The renowned Austrian physicist Erwin Schrödinger took up the theme and in March of 1926 showed how wave equations could be used to predict the behavior of electrons. An energy transition, he said, was caused by a change in wave frequency rather than by any kind of quantum leap. The physics community greeted the assertion with enthusiasm, particularly since wave equations were much more familiar beasts and far easier to tackle than Heisenberg's formulas.

In the early 1920s, dissatisfied with current models for the behavior of subatomic particles, German physicist Werner Heisenberg plunged into the new field of quantum mechanics. By 1927, he had developed his "uncertainty principle"—the fundamental tenet of quantum physics that two related properties of a particle cannot be precisely measured at the same time. An observer can determine either a particle's exact position in space or its precise momentum, but never both simultaneously. By declaring that all subatomic behavior is a matter of probabilities, Heisenberg set the stage for quantum theories of the forces controlling that behavior.

GUARANTEED UNCERTAINTY

But the particle aspects of the subatomic world would not be denied, and Heisenberg continued to search for an answer to the discrepancy between the wave and particle descriptions. He had it by 1927. Examining the odd relationship between position and momentum that his formulas indicated, he realized that one could be pinned down only at the expense of the other; the more specific the value calculated for one, the less specific was the value that could be determined for the other. In other words, a very small but irreducible amount of uncertainty was built into the quantum world. A full description of an electron was therefore a matter of probabilities. In this way, an electron could be both particle and wave, because precisely locating its position—that is, defining it as a particle—made it impossible to describe its momentum, and specifying its momentum, or its wavelike nature, rendered its particle-like position completely unpredictable.

As counterintuitive as it seems in the everyday world, this so-called uncertainty principle turned quantum mechanics into a workable, albeit no less perplexing, theory. As Niels Bohr once commented to a student, "If anybody says he can think about quantum problems without getting giddy, that only shows he has not understood the first thing about them." Einstein himself found the world view represented by the notion of uncertainty so incomprehensible that he argued against the concept for many years, eventually admitting only a grudging acceptance. Others, perhaps more willing to suspend their disbelief, saw that quantum mechanics had practical implications for the theory of electromagnetism, whose equations would have to be reformulated to take quantum effects into account. In particular, because photons are electromagnetic radiation and thus cannot exist without an electromagnetic field being present, theorists now had to know how photons and electrons interacted in terms of a quantum theory for electromagnetic fields. In other words, they needed a quantum field theory.

English physicist Paul Dirac came up with the revolutionary concept in 1928, calling it quantum electrodynamics (QED) because it represented the application of quantum theory to the dynamics of electromagnetic fields. By treating the electromagnetic field quantum mechanically, Dirac and others who applied his theory were able to describe the details of various interactions between photons and electrons. Perhaps his biggest coup was incorporating special relativity, with its revelation that matter and energy are equivalent, into an equation of motion that explained the effects of electrons traveling close to the speed of light. Dirac showed that quantum behavior involves the creation and annihilation of particles, and transformations back and forth between energy and matter. For example, an electron moving to a lower-energy quantum state and emitting a photon was redefined as the destruction of an electron and its replacement by a newly created photon and another electron of lower energy.

Dirac's equation, as his theory was also known, held further surprises. For one, it predicted the existence of antimatter, a precise counterpart to matter that is different in one specific aspect. The electron's antimatter partner, for example, is the positron, which differs only in having a positive charge. Matter-antimatter interactions produce a total annihilation, in which both particles disappear in a burst of pure energy—the creation of two highly energized photons. But for those who would ultimately seek to unify the forces, quantum electrodynamics had even more significant news: The electromagnetic force was itself made up of particles.

According to quantum electrodynamics, when two electrons come close enough together to experience each other's influence, they exchange a special class of photon that represents the electromagnetic force causing them to interact. In effect, the photon carries that force between the two electrons. Dirac's equation indicates that these photons operate by means of the uncertainty principle, popping into existence through a so-called quantum fluctuation for so fleeting a time that they are impossible to detect. As a result, and to distinguish them from the familiar observable photons that

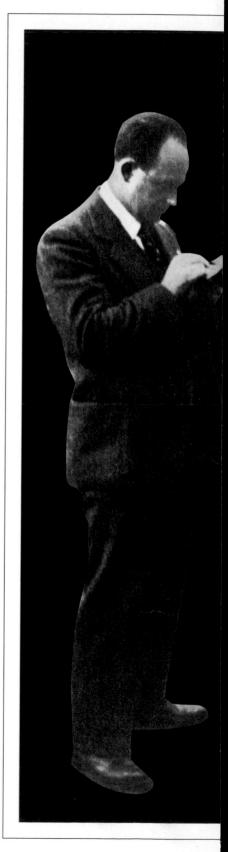

In 1933, Italian physicist Enrico Fermi *(left)* postulated that nuclear decay was governed by the weak interaction, a tenuous force carried by particles that Fermi left unidentified. In 1935, Japanese scientist Hideki Yukawa spelled out the basic details of the strong force that binds nuclear protons. He named his force carriers "mesons."

constitute electromagnetic radiation, they are known as virtual photons.

The notion of force-carrying particles would eventually become a common foundation for descriptions of all the forces. In the meantime, however, QED itself ran into trouble. Heisenberg and Austrian physicist Wolfgang Pauli, who had helped discover electron spin, noticed that some QED calculations yielded infinity as an answer, rendering them meaningless. The difficulty arose in equations intended to determine the interaction of an electron with its own field, a concept known as the electron's self-energy. The seriousness of the problem kept quantum field theory from reaching its full potential for nearly twenty years.

THE WEAK AND THE STRONG

For a brief period not long after the publication of Dirac's QED, some scientists thought they were on the verge of understanding everything there was to know about the atom and that physics would come to an end as a result, all its puzzles solved. The optimism was buoyed by laboratory confirmation of the existence of antimatter in 1932 and the discovery that same year of the neutron, which explained previously unaccounted-for mass in the nucleus. But physics was far from over. Longstanding mysteries that were finally being tackled with some success were about to complicate the picture by introducing two new fundamental forces.

Since the last decade of the nineteenth century, investigators had been studying radioactive minerals composed of atoms that emit highly energized electrons apparently from the nucleus, a process now called beta decay. (Protons are sometimes known as alpha particles and electrons as beta.) How this was happening was not at all clear, particularly when experiments conducted in 1927 showed that the energy emitted in the form of electrons did not equal the energy lost by the nuclei, as it should. To make up the difference, Wolfgang Pauli had suggested in 1930 that another previously unknown particle was also always emitted, although this was little more than argument by necessity and he would not publish it. Toward the end of 1933, however, Italian physicist Enrico Fermi published a theory that not only supported Pauli's tentative supposition but also explained beta decay as the product of a new force—the so-called weak interaction.

Fermi had chosen quantum electrodynamics as his model, despite its problems with infinities. He postulated that beta decay involved the changing of a neutron into a proton through the creation and emission of an electron and an additional, highly energetic but chargeless particle that he dubbed the neutrino. (Later, physicists realized that the chargeless particle must be an antineutrino, which would spin in a direction opposite to that of an electron and thereby cause the two particles' quantum spin values to cancel each other out, as the theory required.) Calculations based on laboratory measurements showed that the force mediating these transformations was extraordinarily weak and could act only over a very short range, in contrast to electromagnetism and gravity, which have infinite ranges. Fermi would not speculate as

to the precise nature of the particles that would carry this force as the photon did electromagnetism, and soon he turned to another endeavor, the work for which he is best known: bombarding atomic nuclei with neutrons, creating radioactive elements, and eventually splitting the atom, thereby ushering in the atomic age. In the end, however, Fermi's weak-force theory was so perfectly executed and matched observed behavior so well that it has survived largely intact to the present day as a useful approximation of what is really happening within atoms during beta decay. And in time, the weak force carrier particles that Fermi had left unidentified in his equations were both predicted and found.

Just as Fermi was developing his theory, other physicists were pondering another odd feature of subatomic behavior. Given their like positive charges, protons in the nucleus of an atom should be repelled instead of somehow sticking together. One early speculation was that the electromagnetic attraction keeping negatively charged electrons bound to nuclei was strong enough to overwhelm the electromagnetic repulsive force within nuclei, but the numbers never added up. By about 1930, many physicists were convinced that some other, unknown fundamental force must be at work, capable of outweighing electromagnetism over the very short range of an atomic nucleus. A complete articulation of this strong nuclear force would take several decades to emerge, but its essential details were spelled out in 1935 by Hideki Yukawa, a brilliant Japanese scientist. Yukawa had taken up the study of theoretical physics a few years earlier, much to the disappointment of his father, who had wanted him to enter the more practical field of geology.

During graduate studies at Kyoto University, Yukawa had been exposed to quantum electrodynamics and had spent a fair amount of time struggling with the infinities that marred the equations. He turned to the atomic nucleus primarily out of frustration at his inability to solve QED's problem. But he did not venture all that far from the theory in his new line of investigation into what was holding protons and neutrons together in the nucleus. After spending two years pursuing dead-end hypotheses, and aided by Fermi's postulation of a weak force, he came up with a theory for the nuclear binding action that leaned heavily on the vocabulary and syntax of QED, even while representing a distinctly different phenomenon.

In essence, Yukawa started from the assumption that all forces should be fundamentally similar. If electromagnetism and gravity consisted of fields, he reasoned, then a strong nuclear force would also. And if quantum field theory worked for electromagnetism, it should for the new force as well. Following a line of thinking first developed by Heisenberg, he imagined that neutrons and protons were surrounded by fields consisting of virtual particles that mediated the strong force. His theory showed that inside the nucleus, neutrons were always turning into protons and protons into neutrons—maintaining overall the appropriate number of each for the particular element—by exchanging virtual particles that carried either positive or negative charges. The constant flow of these two types of particles was what

A Particle Family Tree

Nature in all its diversity consists of fewer than twenty basic types of particles, making up both matter and the forces that act upon it. The list of so-called elementary particles, whose main entries are shown at right, has grown dramatically in the past fifty years. Earlier this century, physicists considered the proton, neutron, and electron the most fundamental constituents of matter, and the photon was the only known force carrier. Later discoveries indicated that protons and neutrons are themselves composed of smaller particles, which come in several varieties and combine to create hundreds of additional composite particles. Researchers have identified other indivisible bits of matter and also confirmed the existence of the strong- and weak-force carriers, leaving only the graviton—the hypothetical gravity particle—undetected.

Each particle possesses a unique combination of values for mass, charge, and spin that distinguishes it from all others; each also has an antimatter equivalent (not shown here) that is opposite in one property, usually charge. The general categories into which particles fall are based on shared characteristics. The broadest classes are fermions *(light blue panels)*, which have fractional spin values such as 1/2 or 3/2, and bosons *(gray panels)*, which have integer spins of 0, 1, or 2. All of the force carriers are bosons, and all but a few matter particles are fermions.

The fermions themselves come in two forms—leptons and quarks. Leptons are alike in not responding

LEPTONS

Electron
Mass   1
Charge −1
Spin   ½

Electron Neutrino
Mass   ?
Charge 0
Spin   ½

Muon
Mass 207
Charge −1
Spin   ½

Muon Neutrino
Mass   ?
Charge 0
Spin   ½

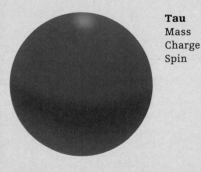

Tau
Mass 3,490
Charge   −1
Spin   ½

Tau Neutrino
Mass   ?
Charge 0
Spin   ½

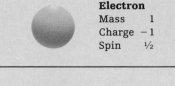

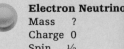

FERMIONS

QUARKS

Up
Mass 10
Charge +2/3
Spin 1/2

Down
Mass 20
Charge −1/3
Spin 1/2

Charm
Mass 2,900
Charge +2/3
Spin 1/2

Strange
Mass 195
Charge −1/3
Spin 1/2

Top
Mass 58,000
Charge +2/3
Spin 1/2

Bottom
Mass 8,900
Charge −1/3
Spin 1/2

BOSONS

FORCES

Photon (Electromagnetism)
Mass 0
Charge 0
Spin 1

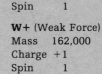

W+ (Weak Force)
Mass 162,000
Charge +1
Spin 1

W− (Weak Force)
Mass 162,000
Charge −1
Spin 1

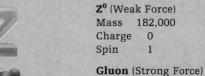

Z⁰ (Weak Force)
Mass 182,000
Charge 0
Spin 1

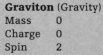

Gluon (Strong Force)
Mass 0
Charge 0
Spin 1

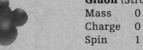

Graviton (Gravity)
Mass 0
Charge 0
Spin 2

BARYONS
(Quark triplets)

Proton
A proton consists of two up quarks *(small wedges)* and one down quark *(larger wedge).* Gluons *(blue),* which carry the strong force, hold the quarks together. Each up quark has a +2/3 charge, the down quark a charge of −1/3; to-gether, they give the proton a charge of +1.

Neutron
A neutron is made of two down quarks *(large wedges)* and one up quark *(small wedge),* bound by gluons *(blue).* Neutrons have no charge be-cause the two −1/3 charges of the down quarks offset the +2/3 charge of the up quark.

MESONS
(Quark doublets)

Kaon
Mesons are made of two quarks, one of which is always an antipar-ticle. In the kaon, a strange quark *(solid red)* combines with an anti–up quark *(red outline).* As with baryons, gluons bind the particle.

Pion
A pion consists of an up quark *(sol-id red)* bound by gluons to an anti–down quark *(red outline).* Many other types of mesons exist, each fashioned from a different pair of quark and antiquark.

held the neutrons and protons together; the flow was itself the strong force.

Yukawa's crowning achievement was elucidating the nature of the virtual particles. Since the strong force operated over an extremely short range—less than a trillionth of a centimeter, or about the diameter of a proton—its carriers had to decay extremely quickly; otherwise, the force would extend farther and would interfere with the electromagnetic force at work between an atom's protons and electrons. Through a complex series of calculations, Yukawa was able to compute a mass for these particles, estimated at about 200 times that of the electron and a tenth that of the proton. They were eventually called mesons as a result, from the Greek word for middle. Yukawa further postulated that meson decay was triggered by the weak force, which itself must have similar types of virtual particles that break down into the

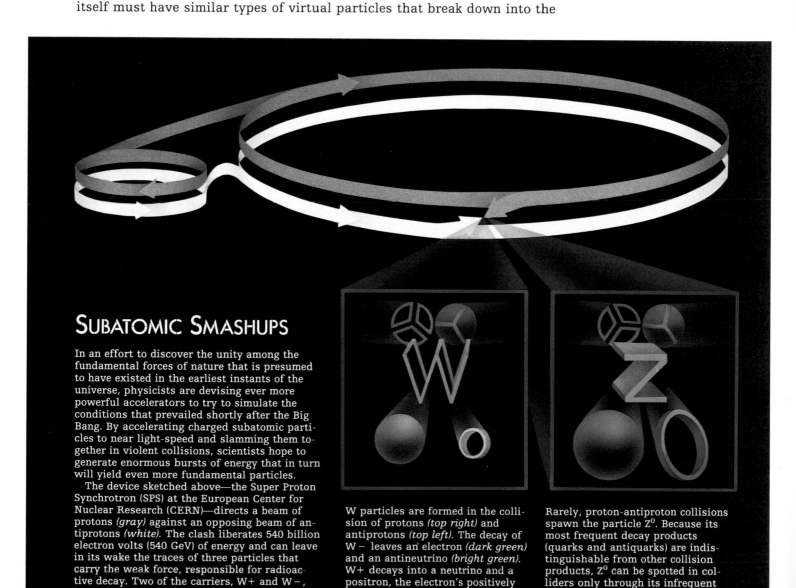

SUBATOMIC SMASHUPS

In an effort to discover the unity among the fundamental forces of nature that is presumed to have existed in the earliest instants of the universe, physicists are devising ever more powerful accelerators to try to simulate the conditions that prevailed shortly after the Big Bang. By accelerating charged subatomic particles to near light-speed and slamming them together in violent collisions, scientists hope to generate enormous bursts of energy that in turn will yield even more fundamental particles.

The device sketched above—the Super Proton Synchrotron (SPS) at the European Center for Nuclear Research (CERN)—directs a beam of protons (gray) against an opposing beam of antiprotons (white). The clash liberates 540 billion electron volts (540 GeV) of energy and can leave in its wake the traces of three particles that carry the weak force, responsible for radioactive decay. Two of the carriers, W+ and W−, have the same mass but opposite charges; one, Z^0, has a larger mass and no charge.

W particles are formed in the collision of protons (top right) and antiprotons (top left). The decay of W− leaves an electron (dark green) and an antineutrino (bright green). W+ decays into a neutrino and a positron, the electron's positively charged antiparticle.

Rarely, proton-antiproton collisions spawn the particle Z^0. Because its most frequent decay products (quarks and antiquarks) are indistinguishable from other collision products, Z^0 can be spotted in colliders only through its infrequent decay into particles such as the muon and antimuon shown here.

electrons and antineutrinos of beta decay. Almost as an afterthought, he added that although the hypothesized mesons would disappear too quickly to be observed in their role as force carriers, they might temporarily be detectable as random products of highly energetic subatomic collisions.

Such collisions occur naturally throughout the atmosphere because of cosmic rays—streams of charged particles that continually rain down on Earth from unknown sources in space, striking air molecules and producing cascades of all sorts of energized particles. In 1932, American physicist Carl Anderson had taken advantage of cosmic rays to discover the positron, the electron's antimatter counterpart, using a device known as a cloud chamber, a sealed container filled with moist air that indicated the paths of speeding particles with trails of condensation. In 1936, Anderson studied another

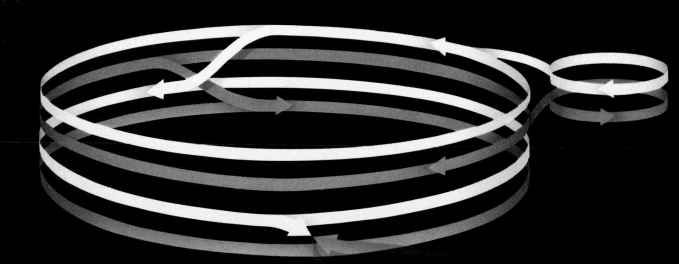

UPPING THE ANTE

The highest experimental energy levels have been achieved at the Fermi National Accelerator Laboratory (FermiLab) near Chicago. Like the SPS at CERN, this complex *(above)* directs protons against antiprotons after drawing the beams from a particle generator *(right)* and running them thousands of times through a loop that uses magnetic fields as a spur. A lower loop equipped with superconducting magnets then boosts the beams even closer to the speed of light, producing an encounter that liberates nearly two trillion electron volts (TeV) and earns the device the title Tevatron.

With the Tevatron, FermiLab scientists hope to detect the top quark, the most massive of six such essential units of matter *(page 27)* and the only one that has yet to be positively identified. Because of its mass, the top quark will emerge only in collisions that produce very high-energy gluons—carriers of the strong force that binds nuclear particles.

The annihilation of the three quarks making up the proton and the three antiquarks that form the antiproton unleashes gluons *(blue)*, which then interact to engender particles such as the massive top quark and anti–top quark.

strange particle, which was negatively charged like an electron but nearly 200 times more massive, the mark that Yukawa had predicted for the meson. Yukawa's 1935 paper, which had been largely ignored, suddenly received fresh attention, and several noted physicists were soon proclaiming the discovery of the strong-force carrier.

But they were wrong. It turned out that this first particle, now known as a mu meson, or muon, did not act precisely the way it was supposed to. If muons were the carriers of the strong force, then a negative muon passing near an atomic nucleus would feel the combined effects of the strong force and of electromagnetic attraction to the positively charged protons in the nucleus and would plunge immediately into the nucleus, before getting a chance to decay. But experiments conducted toward the end of World War II showed

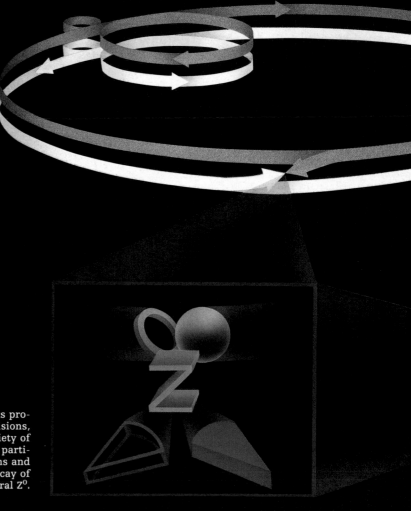

A Better Z⁰-Trap

Although proton-antiproton collisions occasionally produce the elusive weak-force carrier Z^0, the particle's traces are usually hard to pick out of the welter of collisional and decay products characteristic of these encounters. The best way to spy Z^0's trail is to smash electrons and positrons in a burst of energy equivalent to its mass—about 91 GeV. This level is tough to achieve, however, because electrons and positrons radiate away considerable energy in rounding curves. Researchers at CERN have thus devised the Large Electron-Positron Collider *(near right)*, whose circumference of sixteen miles greatly reduces radiation loss. Boosted initially by the SPS there *(inner circle)*, the electron and positron beams peak at 45.5 GeV, for a combined charge on collision that produces Z^0s in abundance.

Because research into more massive particles would require higher-energy collisions and an even larger and more costly circular track, linear accelerators are regarded as the most feasible alternative for the future. The Stanford Linear Collider, or SLC *(far right)*, guides electrons and positrons down a two-mile-long accelerator toward collision in an oval.

Unlike the quarks and antiquarks produced in proton-antiproton collisions, which may be the result of a variety of processes, the presence of those particles in collisions between electrons and positrons is a sure sign of the decay of the ephemeral Z^0.

that this did not happen, and the muon was rejected as a candidate for carrier of the strong force (it turned out not even to belong to the meson class). Finally, in 1946, a heavier particle was discovered, the pi meson, or pion, and it reacted in the anticipated fashion: Yukawa at last had confirmation of his theory.

QED'S TRIUMPH

After the war, the investigation of the four fundamental forces underwent a significant shift. Although all the forces were now at least broadly sketched, they were missing details, and theorists worked at fine-tuning descriptions and probing for connections. Meanwhile, experimental physicists, who were learning to manufacture their own high-energy collisions in particle accelerators *(pages 28-30)*, were coming up with a host of new particles, which both confirmed existing hypotheses and inspired new ones. Along the way, compelling cases were being made for unification, at least among the three quantum field theories. These cases all ultimately depended on a brilliant insight, achieved independently in 1948 by three physicists, that finally cured quantum electrodynamics of its infinity woes.

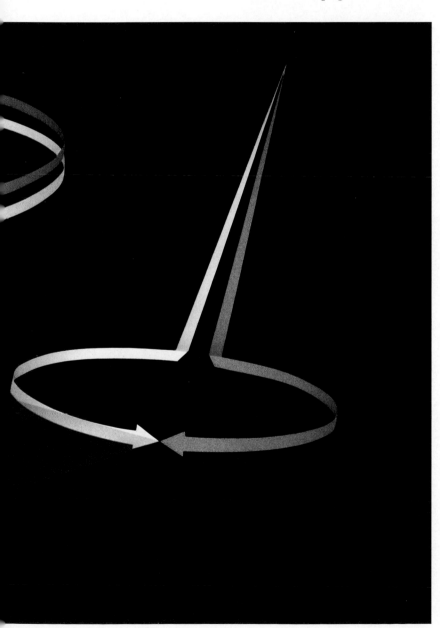

The three men, who shared a Nobel Prize awarded in 1965, were Americans Julian S. Schwinger and Richard P. Feynman, and Shin'ichirō Tomonaga from Tokyo. Their technique, which evolved from an idea developed in the 1930s by several other scientists, was known as renormalization. It amounted to a mathematical ploy that, in effect, supplied an infinity on one side of an equation to compensate for an infinity that had cropped up on the other. Feynman's description of how this was accomplished may have been the most comprehensible. He explained that the observed mass of an electron includes both its self-energy (the troublesome mathematical infinity resulting when a particle interacts with its own field) and a quantity known as bare mass, which cannot be observed because of the screening action of the field of virtual particles surrounding the electron like a cloud. Since it was unobservable, bare mass could be assigned an infinite negative value to offset the positive infinity in the calculations of self-energy. This mathematical legerdemain allowed physicists to make accurate predictions about electron behavior. For example, with renormalization, laboratory tests of the strength of an electron's magnetic moment—the magnetic field created by its spin motion around its axis—differed from predictions by only a few parts in 10 billion, an acceptable reduction of a discrepancy that had been much larger under Dirac's original formulation.

The stage was now set for serious attempts to unify the forces. A difficult but inspiring concept, first developed in 1918 but largely overlooked at the time, would help realize these efforts. The concept was symmetry, and it was articulated in relation to modern quantum field theory in 1954 by Chen Ning Yang and Robert Mills, two scientists at the Brookhaven National Laboratory

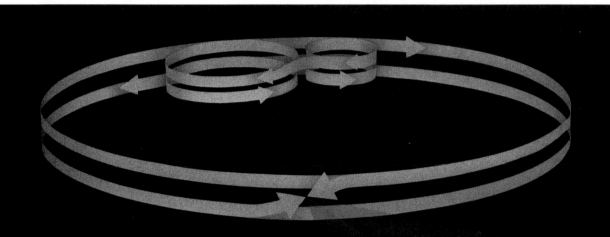

Near the Limit

In 1999, near Dallas, Texas, the Superconducting Supercollider (SSC) is scheduled to make its long-awaited debut. Two opposing beams of protons will be drawn through smaller loops *(top)* and then into the SSC's main ring, nearly fifty-four miles around and equipped with superconducting magnets. The gradual arc of the big ring will minimize the energy lost as protons travel the curved track. After completing three million laps, the protons will collide with a combined force of forty TeV—simulating conditions in the cosmos less than one-trillionth of a second after the Big Bang.

In that explosion, clues may be found to the mystery of how matter emerged from pure energy at the dawn of the universe. For one, physicists hope to detect the so-called Higgs particle *(right)*, carrier of a hypothetical fifth force that confers mass on all particles.

In this scenario, W− and W+ particles generated by a very energetic proton-proton collision *(top)* annihilate each other, creating the hypothetical Higgs particle that lent them mass.

in New York. In simplified terms, symmetry is the maintenance of a certain measurable value in the face of some kind of alteration. As a graduate student at the University of Chicago in the late 1940s, Yang had realized that electromagnetism could be described in terms of symmetry—the velocity of light stays the same regardless of changes in motion. This principle was also known as gauge invariance, gauge being another word for a standard of measurement. The electromagnetic field, then, could be understood as being gauge invariant: The rules governing its behavior did not change over the extent of its range, and symmetry was preserved.

Working with Mills several years later, Yang developed equations that spelled out how gauge field theory could be applied to the strong nuclear force. The result was a new method of approaching the subject, in which the starting point was identifying a symmetry and then mathematically building a gauge field that would maintain it. From these general principles, a cocksure American physicist named Murray Gell-Mann changed the picture of the strong force and the very nature of its particles, taking the whole concept to a new and deeper level.

QUARKS, BY FLAVOR AND COLOR

In particle accelerator and other experiments during the 1950s, researchers came upon many new types of particles—lambdas, sigmas, xis, and others designated by various letters of the Greek alphabet—that were different from one another in some fundamental regard, such as mass or charge, but alike in responding to the strong force. Gell-Mann was determined to find an explanation not only for the distinctions but also for the similarity among these particles, collectively known as hadrons. Gauge theory turned out to hold the key. In 1961, Gell-Mann proposed that all hadrons, which also include the familiar proton and neutron, could be schematically organized in such a way as to maintain an underlying symmetry despite variations in specific particle characteristics. As so often seems to happen, another scientist, Israeli physicist Yuval Ne'eman, reached the same conclusion independently at about the same time. Such a coincidence struck again three years later, when Gell-Mann and U.S. physicist George Zweig separately realized that the symmetry arrangement made sense if particles affected by the strong force were actually composed of different combinations of smaller particles. Zweig called these new elementary particles aces, but Gell-Mann's term, quarks—whimsically borrowed from a line in James Joyce's almost impenetrable novel, *Finnegans Wake*—won out.

Suddenly physicists were looking at an entirely different conception of a basic category of matter, one that quickly grew quite intricate. Carrying on with his fanciful language, Gell-Mann suggested that quarks come in three flavors, arbitrarily labeled up, down, and strange, and that ordinary protons and neutrons consist of different trios of ups and downs. Strange quarks appeared only in unusual particles produced in high-energy experiments, and some hadrons, such as pions, were made up of only two quarks instead of three. Eventually, other researchers would predict three more flavors—charm, top, and bottom—that explained other particle types. To date, only top quarks have not been confirmed experimentally.

But the quark story was still not complete. Working with an enthusiastic young German physicist named Harald Fritzsch, Gell-Mann came up with an important additional wrinkle in 1971. In order to understand properly how quarks combined, it was necessary to imbue them with a further quality, called color, which was a kind of charge through which the strong force acted, similar to the electric charge in electromagnetism. In fact, what Gell-Mann was constructing was an equivalent to quantum electrodynamics for the strong force. Following through on the metaphor of color, he called the theory quantum chromodynamics.

Unlike electric charge, color charge comes in three forms, termed red, green, and blue, and each flavor of quark can carry any one of these colors. The basic rule of interaction is that quarks must form combinations that, in effect, neutralize their separate color charges, just as different colors blend to produce white; this is necessary because the property of color has never been observed and is presumed to be a kind of hidden, unknowable feature. But the biggest alteration to the concept of the strong force came because the theory had moved down a level. Pions, which had been thought to be the fundamental carriers of the strong force, could not be after all, since they—like all other hadrons—are composed of quarks. The strong force holding the quarks themselves together was now considered to be transmitted by a class of entities called gluons. Moreover, there had to be eight different kinds of gluons to accommodate all the various types of quark combinations.

Clearly, electromagnetism and the strong force differed in a number of distinct ways, and although quantum field theory was proving its worth as a general framework, the strong force did not appear to offer a very promising route to unification. Luckily, the same could not be said of investigations that other theorists had been pursuing at about the same time. The path to unity, it seemed, was by way of the weak force.

NEW COMBINATIONS

As with the strong force, gauge theory yielded an improved understanding of the weak interaction, but it did so indirectly. In 1956, two years after he and Robert Mills had published their initial work on symmetry, Chen Ning Yang and another colleague, Tsung Dao Lee, had found a strange breakdown of symmetry in the effect of the weak force on a particle known as the kaon (pages 44-45). This failure of symmetry eventually led another trio of physicists to discover something about the relationship between the forces.

The first of the three to announce a breakthrough was Sheldon Glashow. In the late 1950s, as a postgraduate student at the Niels Bohr Institute of Theoretical Physics in Copenhagen, Glashow employed standard gauge principles—in which symmetries are maintained—to posit a unified theory of electromagnetism and the weak force. Using existing categorizations in which

In 1964, physicist George Zweig of CERN (near right) and CalTech's Murray Gell-Mann independently came up with the idea that particles such as the proton were actually composed of even smaller particles. Zweig called his objects "aces," but Gell-Mann's name—"quarks," from a line in the James Joyce novel Finnegans Wake—stuck. In work that later earned a Nobel prize, Gell-Mann went on to propose three "flavors" of quarks (physicists have since confirmed five and theorize a sixth). He also suggested that quarks have "color," a property similar to charge but based on spin.

matter particles are called fermions and force particles bosons, Glashow predicted the existence of three weak-force bosons that were analogous to the photon. Two of them, dubbed W+ and W−, carried charge to account for the fact that beta decay sometimes transmitted a negative and sometimes a positive charge; one of them, the Z, was neutral. Glashow's bold attempt, which would ultimately prove accurate, failed for the time being in one key, familiar regard: His equations produced infinite results. This forced him to renormalize, but the procedure worked only if the carriers for the unified electroweak force were all massless, like the photon. Unfortunately, complex energy-to-mass conversions indicated that weak-force bosons had to be especially massive, on the order of 100 times more massive than protons.

THEN THERE WERE THREE

The dilemma was not easily or quickly resolved, and depended on contributions from Abdus Salam of Pakistan, and Glashow's high-school friend and college classmate at Cornell, Steven Weinberg, the two other physicists who would ultimately share a Nobel prize with Glashow for linking electromagnetism and the weak force in a so-called electroweak theory. The clinching factor turned out to be broken symmetry. Building on concepts and equations developed by several other theorists—most notably Jeffrey Goldstone of Cambridge and Peter Higgs of the University of Edinburgh—Salam and Weinberg showed that photons and weak-force bosons could be considered as different manifestations of what had, early in the universe's history, been the same fundamental entity, and that they had become distinguishable by the breaking of a previously intact symmetry. The symmetry breaks allowed massless bosons to pick up mass.

Renormalization continued to be somewhat problematic in versions of this theory presented by Salam and Weinberg in 1967, but in 1971, Dutch physicist

Gerard 't Hooft devised an ingenious mathematical technique, akin to earlier procedures but with some added refinements, that banished infinities forever. More than ten years later, particle accelerators provided a final confirmation by detecting W and Z bosons with almost 100 times the mass of a proton—just as had been predicted by the electroweak theory.

The next step seemed virtually inevitable. Despite observable major differences in behavior, scientists thought that all three of the forces described by quantum field theories ought at some level to be identical. The chore, which was first tackled in

In 1979, Sheldon Glashow *(near right)*, Abdus Salam *(far right)*, and Steven Weinberg *(below)* shared the Nobel prize for a theory unifying the electromagnetic and weak forces. Glashow's initial work in the late 1950s ran into problems of mass: One set of equations called for carriers of the electroweak force to be massless, although they were known to be massive. In 1967, Salam and Weinberg showed that shortly after the Big Bang, a symmetry break allowed massless force carriers to acquire mass.

the early 1970s, was somehow to bring the strong nuclear force and the new electroweak force into accord through what the searchers were calling a grand unification theory, or GUT.

By 1974, Glashow and his Harvard colleague Howard Georgi were claiming success. They had found a GUT in which gluons were indistinguishable from photons and the weak-force bosons, and electrons, which normally do not respond to the strong force, acted just like quarks, which are affected. The specifics of their theory were intimately related to the complicated mathematics of gauge and quantum field theories, but the key revelation was that under conditions of extremely high energy, the various particles and forces became interchangeable.

Cosmologists were ecstatic. The new GUT seemed to fit with the prevailing theory of the Big Bang, the hypothesized explosion that brought the whole universe—matter, energy, time, and space—into existence some 15 or 20 billion years ago. An infinitesimal fraction of a second after the instant of the blast, according to the cosmologists, all of space was suffused with as much energy as Glashow and Georgi were saying was necessary to unite the forces. An equally tiny fraction of a second later, cooling of the expanding fireball caused the strong and electroweak forces to condense out of their primordial union and begin to affect different particles in different ways. In a similar manner, electromagnetism and the weak force became distinct.

At first there seemed no possibility of testing the theory, because no accelerator could ever duplicate the trillion-degree temperatures of the nascent cosmos. But Georgi and Glashow's theory did have one testable implication: It predicted that protons—once thought to be eternally stable—should eventually decay and their quarks come apart. Researchers soon began to set up experiments designed to detect such an event, which was bound to be exceedingly rare. To date, no proton has been caught in the act of decay, a failure that seriously challenges at least this one version of a grand unification theory. Theorists remain confident, however, that some kind of three-force union will ultimately hold up.

THE MISSING PLAYER

Despite the ambitious title, grand unification theories suffer from one glaring omission: They completely ignore gravity. And so it has been since the early years of the century. After his seminal work on relativity, Einstein spent many fruitless years searching for a unifying connection between gravity and electromagnetism. Later, in light of the stunning formulations of quantum field theory, other investigators blithely averred that the gravitational field must have a quantum particle associated with it. They even gave it a name, the graviton. But by no stretch of mathematical skill could they get a quantum theory of gravity to work. Unification had, in a sense, reached an impasse. Before a true Theory of Everything could emerge, physicists would have to look at all of the forces from fundamentally different perspectives.

THE ORIGIN OF MATTER

Second only to the mystery of why the universe came into existence is the puzzle of how the matter in it managed to last more than a minuscule fraction of a second. Physicists have long known that all the subatomic building blocks of matter have antimatter counterparts, particles identical in mass but opposite in charge and other characteristics. When the high-energy conditions of the early universe are approached in particle-accelerator experiments, both types of particles leap into existence in precisely equal numbers—created in consonance with Einstein's law of the interchangeability of matter (or antimatter) and energy. The particle opposites are self-destructive by nature: In theory and in fact, contact between a particle and its antiparticle results in mutual annihilation in a blaze of energy. Yet the Big Bang deviated from this scenario. If the creation of particles and antiparticles had proceeded on a perfectly equal basis in the primordial forge of the cosmos, nothing would have survived to form the stars, dust clouds, planets, and other objects seen in the heavens today.

A clue to the riddle has been found in a certain lack of impartiality in one of the fundamental forces of nature—the so-called weak nuclear force, which regulates a number of particle interactions, including certain forms of decay. The clue expresses itself in the behavior of the particles arrayed at right, quick-change artists governed by the exotic rules of quantum mechanics. As explained on the following pages, efforts to fathom their ways have not always been comfortable for scientists. Some cherished beliefs about the nature of physical laws have been overthrown, and pieces are still missing from the puzzle of why the cosmos is an immense vessel for matter rather than an ancient vacancy.

Operating only at close range, the weak force interacts with two related families of subatomic particles called leptons *(right)* and quarks *(far right)*. The generic lepton *(top right)* represents the members of this family as they existed in the earliest instants of the universe. Other leptons include electrons and muons, and their antimatter counterparts, as well as associated types of neutrinos and antineutrinos, represented here by one type for simplicity.

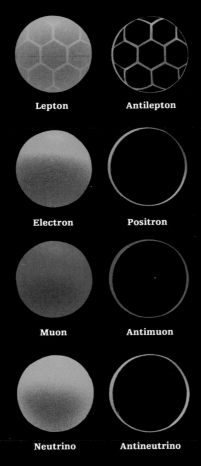

Lepton   **Antilepton**

Electron   **Positron**

Muon   **Antimuon**

Neutrino   **Antineutrino**

Quarks and their corresponding antiquarks *(top right)*—both available in six different varieties, or "flavors," known as up, down, strange, charm, bottom, and top—join together to produce a large number of composite particles. Here, two up quarks and one down quark have assembled to create a proton *(near right)*; two anti–up quarks and one anti–down quark yield an antiproton *(far right)*.

The combination of a quark and an antiquark yields particles called mesons, their nature depending on the identity of their constituents. In the top row at right, a down quark and an anti–strange quark form a neutral kaon, or k-meson; an anti–down and a strange quark yield its antiparticle. In the bottom row, an up and an anti–down quark produce a positive pion (pi-meson), and a down and an anti–up quark form a negative pion.

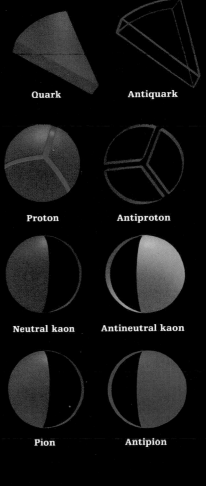

Quark   **Antiquark**

Proton   **Antiproton**

Neutral kaon   **Antineutral kaon**

Pion   **Antipion**

Four types of messenger particles mediate all interactions between other particles. Each is associated with one of the four basic forces of nature. The intermediate vector boson *(top left)* is the agency of the weak force; the graviton *(top right)* accounts for the interactions observed as gravity; the photon *(bottom left)* is responsible for electromagnetism; and the gluon *(bottom right)* transmits the strong force that holds quarks together in composite particles.

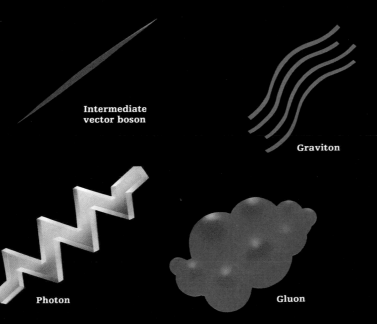

Intermediate vector boson

Graviton

Photon

Gluon

In the ultra-energetic conditions immediately after the Big Bang, a species of very massive and very short lived particle known as the X Higgs boson was created (shown below with its antiparticle). It quickly decayed into quarks and leptons.

X Higgs boson   **Anti–X Higgs boson**

QUESTIONS OF SYMMETRY

In describing the deepest attributes of nature, phys-icists speak of various forms of symmetry, a word derived from the Greek for "same measure." In the familiar world, a sphere offers an example of a high degree of symmetry: It looks identical from every an-gle. Similarly, many of the workings of nature, such as gravity or electrical effects, are independent of par-ticular points of reference; they do not distinguish between places, times, and directions but instead work the same way everywhere and always. This has momentous implications. From nature's symmetries arise conservation laws that govern physical interac-tions—laws, for instance, asserting that energy, linear momentum, and angular momentum are always con-served, never increased or decreased.

One important type of symmetry is called parity. In everyday terms, it is somewhat akin to saying that there is no incompatibility between an event and its reflection in a mirror, which reverses left and right. That is, a physical phenomenon and its mirror image are equally possible. Since the 1930s, scientists have known that the operations of gravity, electromagne-tism, and the strong nuclear force do not differentiate between left and right. Naturally enough, it was as-sumed that parity would also hold true for the least familiar of the four basic forces of nature, the weak force, which controls particle decay. Thus, the decay of a muon particle was thought to occur in the manner shown here and explained at far right. But this was only an assumption, tidy, reassuring—and unproven.

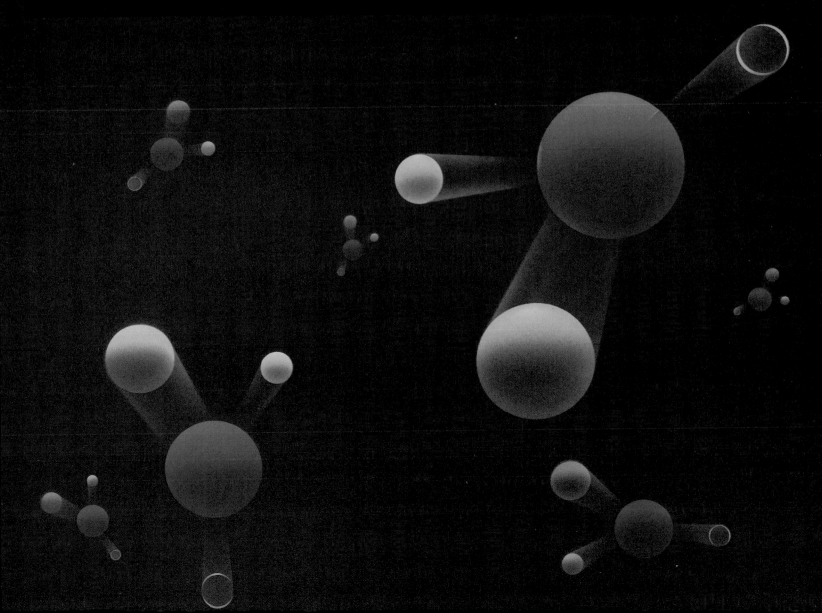

An important property of fundamental particles is spin—equivalent to rotation about an axis.
Quarks, electrons, neutrinos, and
several other particles are described as left-handed or right-
handed, depending on the relative
orientation of their spin and their
direction of linear motion. (In this
convention, the thumb of the hand
is aligned with the direction of linear motion and the fingers with the
direction of spin.) At near right, a
muon *(blue)* is shown decaying into
a neutrino *(bright green ball)*, an
antineutrino *(bright green ring)*,
and a left-handed electron *(dark
green)*. At far right, a muon yields a
right-handed electron. If muon decay observed parity, equal numbers
of right- and left-handed electrons
would be produced *(background
illustration)*.

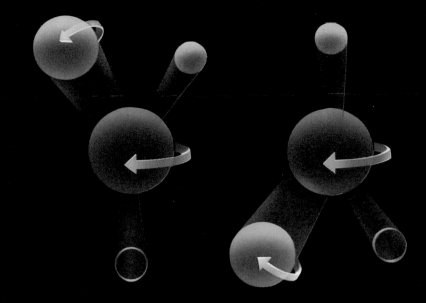

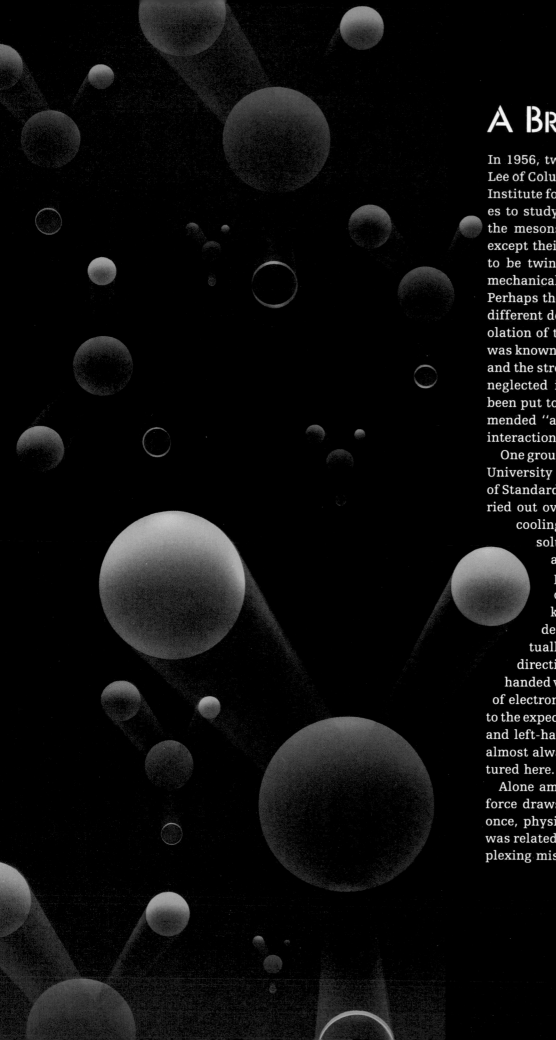

A BREACH OF PARITY

In 1956, two Chinese émigrés, physicists Tsung Dao Lee of Columbia University and Chen Ning Yang of the Institute for Advanced Study in Princeton, joined forces to study a baffling pair of fundamental particles, the mesons known as tau and theta. In every way except their manner of decay, the particles appeared to be twins. Lee and Yang pondered the quantum-mechanical issues involved and made a bold guess: Perhaps the particles were in fact identical and their different decay modes resulted from a weak-force violation of the form of symmetry called parity. Parity was known to hold true for gravity, electromagnetism, and the strong force, but the weak force—obscure and neglected in comparison to the others—had never been put to the test. The researchers urgently recommended "an experiment to determine whether weak interactions differentiate the right from the left."

One group, headed by Chien-Shiung Wu of Columbia University and Ernest Ambler of the National Bureau of Standards, designed an exquisite trial that was carried out over a period of six months. It consisted of cooling atoms of radioactive cobalt to near absolute zero, lining up their nuclear spins with a magnetic field, and recording the flashes produced when electrons emitted by the decay of neutrons—the radioactive process known as beta decay—struck crystalline detectors. The results were unequivocal. Virtually all of the electrons were emitted in a direction indicating that they were of the left-handed variety. The same was also shown to be true of electrons emitted when muons decayed. Contrary to the expected yield of equal numbers of right-handed and left-handed electrons *(pages 40-41)*, muon decay almost always produces the left-handed kind, as pictured here.

Alone among the four basic forces, then, the weak force draws a distinction between left and right. At once, physicists began to wonder if this distinction was related to another asymmetry in nature—the perplexing mismatch of matter and antimatter.

Symmetry Restored

The discovery that the weak force violates parity rocked the world of physics—and swiftly won Lee and Yang the Nobel prize. Another Nobel laureate, the physicist I. I. Rabi, expressed the prevailing reaction: "In a certain sense, a rather complete theoretical structure has been shattered at its base, and we are not sure how the pieces will be put together."

At once, physicists began searching for a way to restore solidity to the foundations of their discipline. An answer was soon forthcoming—from Lee and Yang, among others. It involved combining two symmetries: the mirror reversal of parity, and a very different axiom asserting that matter and antimatter react identically to gravity, electromagnetism, and the strong force. The theoretical damage done by parity violation had undercut this symmetry as well, again, only for the weak force. But physicists now realized that the two symmetries compensated for each other: If particles are replaced by their antiparticles in a weak-force interaction, the relative orientation of spin and linear motion—handedness, in other words—works in reverse. At right, for example, the decay of antimuons *(blue rings)* yields positrons *(dark green rings)*, the antiparticles of electrons, that are almost entirely right-handed. In effect, the left-handed asymmetry of muon decay is perfectly offset by the right-handed asymmetry of antimuon decay. Nature becomes orderly once again, exercising the weak force without preference for left or right.

This insight was greeted with general relief, but it left physicists without a line of attack on their old dilemma of why the universe contains mostly matter. If dual symmetry restored even-handedness to nature, how could an imbalance have come about?

A New Flaw in Nature's Mirror

For seven years after the dual-symmetry idea was proposed, the books of the weak force remained in balance. Then a new tilt came to light. In 1964, James Cronin and Val Fitch of Princeton took a close look at the decay of a subatomic particle called the neutral kaon—an entity that exemplifies why the quantum world makes no sense in everyday terms.

For starters, there are actually two neutral kaons, which can be designated K^0 (pronounced K-zero) and its antiparticle, $\bar{K}^0$ (K-zero-bar); they can also be designated K_1 and K_2. The quantum convolutions arise because K_1 is formed by adding the characteristics of both K^0 and $\bar{K}^0$; according to the rules of quantum mechanics, this makes K_1 "even." K_2 is formed by subtracting $\bar{K}^0$ characteristics from K^0, which makes K_2 "odd." Because K_1 is "even," one set of its possible decay products is a pair of pi-mesons, or pions; because K_2 is "odd," it decays into three pions when it is not decaying into other sets of particles. In their experiments, Cronin and Fitch found that two times out of a thousand decays, K_2 decayed into only two pions instead of the usual three—a tiny but irrefutable anomaly. Since K_2, by definition, must be odd, the researchers realized that it was transforming into another particle, which they named K-long *(box, right)*.

Further investigation, by Cronin and Fitch as well as by other teams of researchers, showed that K-long displayed a preference for decaying into positrons rather than into electrons, direct evidence that symmetry was broken. Said Fitch years later, "This discovery was not expected, it was not wanted when found, and the early explorations were directed to find a way through and around it."

No such way was discovered. The neutral kaon's positron bias is a proven fact. So far, it is the sole indication of a flaw in the combined symmetry of parity and particle/antiparticle interchange. Once again, the universe has demonstrated a capacity for a fundamental lopsidedness in particle interactions—a bias that may be traceable back to the unimaginably energetic conditions of the Big Bang.

The neutral kaon can be either the particle called K^0 *(near right)* or its antiparticle, $\bar{K}^0$ *(far right)*. The weak force changes one into the other so readily that any population of neutral kaons quickly becomes a fifty-fifty mix of the two.

Both K^0 and $\bar{K}^0$ consist of equal amounts of K_1 and K_2, each half-particle, half-antiparticle. The very short lived K_1 decays into, among other things, pairs of pions *(right)*. K_2, in contrast, survives long enough to transform into another particle, known as K-long.

In quantum terms, K-long has a K^0 component slightly larger than its $\bar{K}^0$ component. As a result, the assorted products of its decay will include a positron *(dark green ring)* slightly more often than they will an electron *(dark green ball, inset)*.

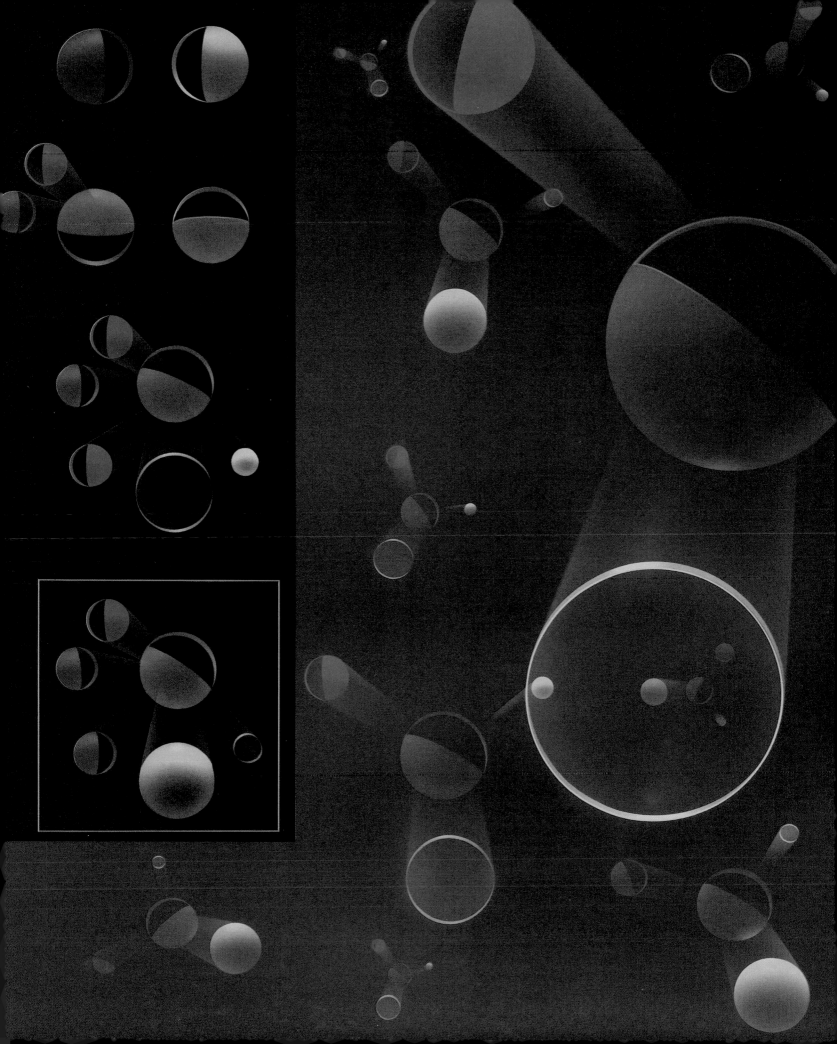

Matter in the Making

The asymmetry that physicists glimpsed in kaon decay may hold the key to favoritism between matter and antimatter during the formative moments of the universe, but the mechanism of that early preference remains murky. It required an exception to the usual rules of symmetry as well as a failure to conserve, or maintain, quantities called quark number and lepton number—defined, respectively, as the difference between the number of quarks and antiquarks or leptons and antileptons in any collection of particles. Physicists believe that the quark and lepton numbers at the beginning of the cosmos were zero: Particles and antiparticles were present in equal quantities. Moreover, no interaction has ever been observed to change these proportions. But the existence of a matter universe leaves no doubt that conservation of quark and lepton numbers was once violated.

The perpetrators, some theorists believe, were ultramassive particles called X Higgs bosons (named for British physicist Peter Higgs), which could exist only in the exceedingly energetic conditions of the very early cosmos. The X Higgs boson and its antiparticle served as a kind of bridge, turning some of the stupendous latent energy of the universe into mass.

At right is a hypothetical depiction of the cosmos about 10^{-33} second after the Big Bang, when the temperature was 10^{26} degrees Kelvin. By then, the basic forces had lost their initial unity and were carried by distinctive particles—photons, gluons, gravitons, and intermediate vector bosons. Quantum theory suggests that, like the neutral kaon, the X Higgs heavyweights *(green Xs)* showed a very slight bias as they decayed into quarks, antiquarks, leptons, and antileptons. But because of their abundance and massiveness, this tiny bias would have enormous consequences. An excess of just one quark and one lepton for every billion annihilations of those particles and their antiparticles would produce an imbalance of matter over antimatter that would account for the billions of galaxies seen in the universe today.

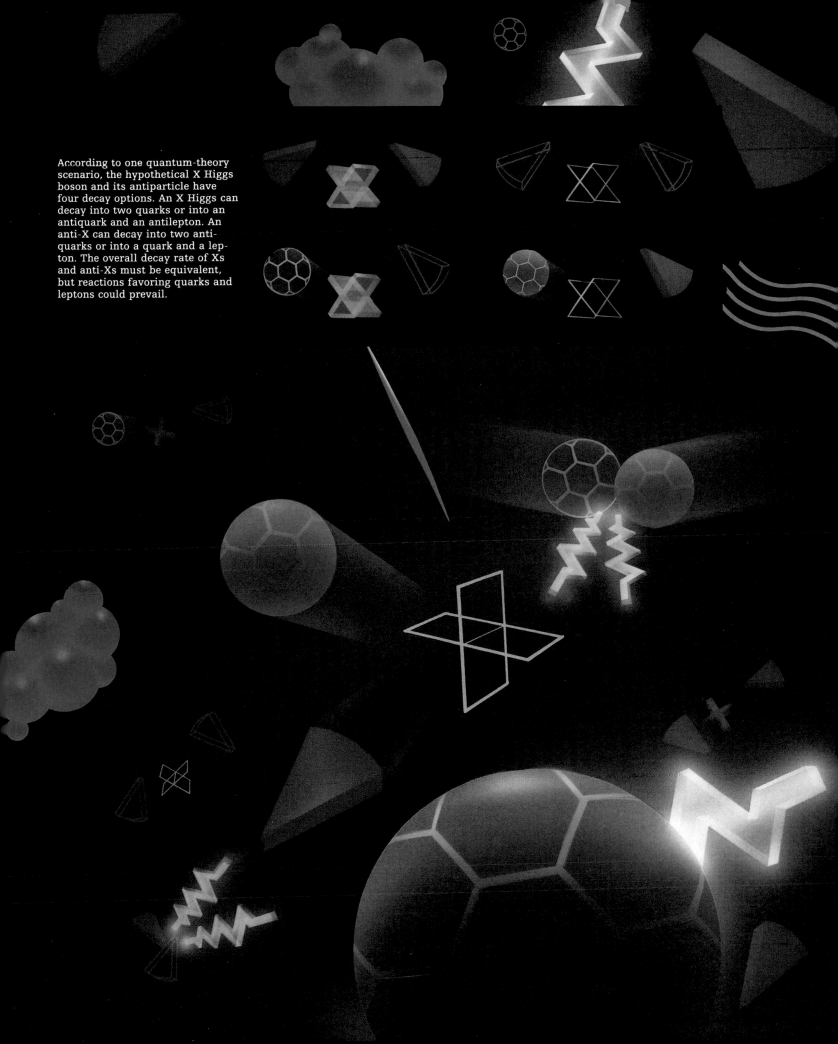

According to one quantum-theory scenario, the hypothetical X Higgs boson and its antiparticle have four decay options. An X Higgs can decay into two quarks or into an antiquark and an antilepton. An anti-X can decay into two anti-quarks or into a quark and a lepton. The overall decay rate of Xs and anti-Xs must be equivalent, but reactions favoring quarks and leptons could prevail.

A Legacy Secured

The period that brought equal numbers of X Higgs bosons and their antiparticles onto the stage of the cosmos is known as the Inflation Era. According to current theory, the volume of the universe very quickly expanded more than a trillion trillion times during this era—a consequence of bizarre quantum effects that caused gravity to work in reverse. By 10^{-33} second after the Big Bang, the explosive phase was over, the universe had dropped to a lower energy state, and the production of the ultramassive X Higgs particles and their antiparticles fell off drastically. Over the next 10^{-6} second, the slightly lopsided decay of the X Higgs breed bequeathed the universe a surplus of quarks (pages 46-47).

Nature proceeded to balance the books as best it could. In the cauldron of the cosmos, quarks and antiquarks joined together in protons, antiprotons, neutrons, and antineutrons—only to engage in an orgy of self-destruction. As rapidly as they were created, the newly made particles annihilated one another in flashes of energy. By one minute after the Big Bang, the process of mutual obliteration had removed all antimatter from the universe. Only a minuscule fraction of the X Higgs endowment of quarks and leptons survived, the one-in-a-billion excess resulting from asymmetric decay. In effect, matter (a mere pittance, compared to what once was) had been frozen into the universe for all time. What was lost is gone forever, but it continues to whisper the tale of its fate in the form of the cosmic background radiation, the electromagnetic echo of the time of annihilation.

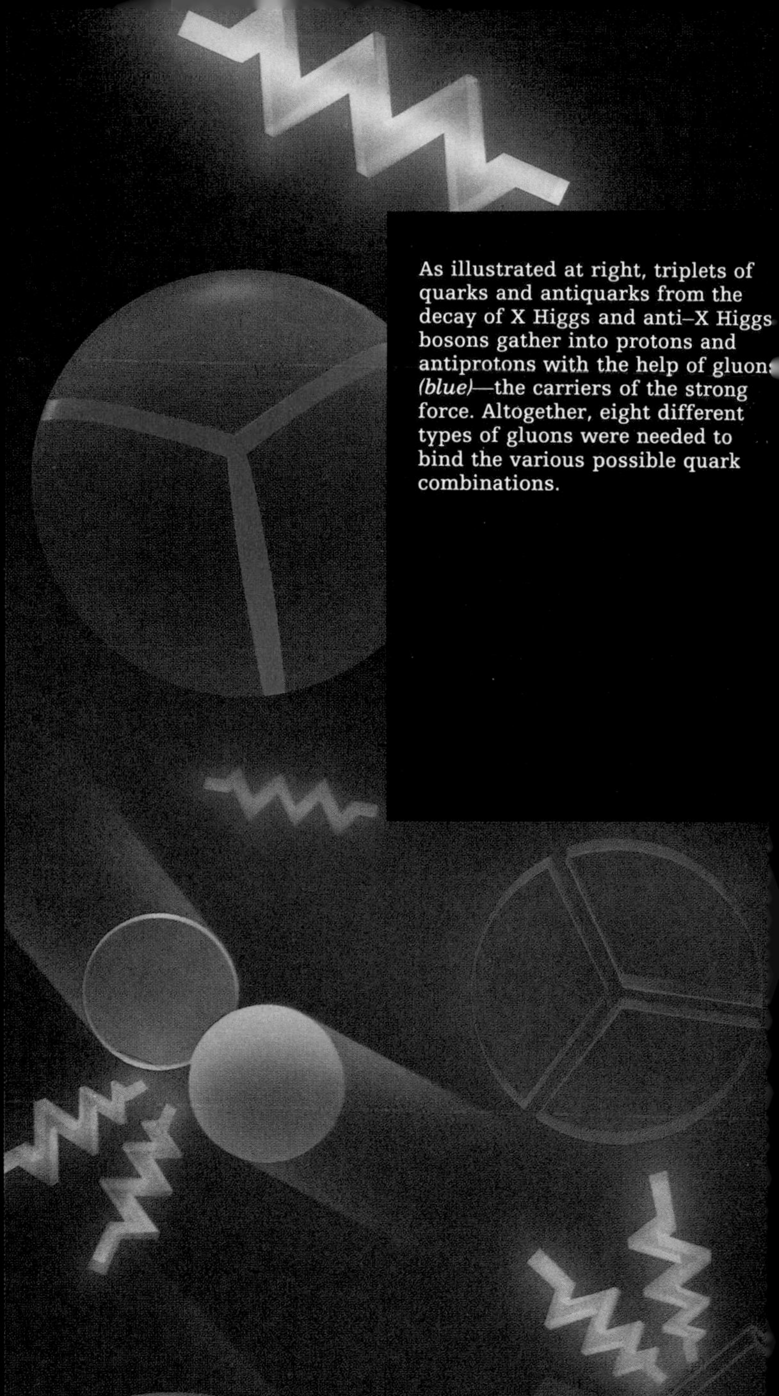

As illustrated at right, triplets of quarks and antiquarks from the decay of X Higgs and anti–X Higgs bosons gather into protons and antiprotons with the help of gluons (blue)—the carriers of the strong force. Altogether, eight different types of gluons were needed to bind the various possible quark combinations.

The meeting of a proton and an antiproton ends in their mutual destruction and the emission of two high-energy photons. In the billions of years since mutual annihilation removed most matter and all antimatter from the universe, cosmic expansion has weakened the relic energy, stretching it from gamma-ray to radio wavelengths.

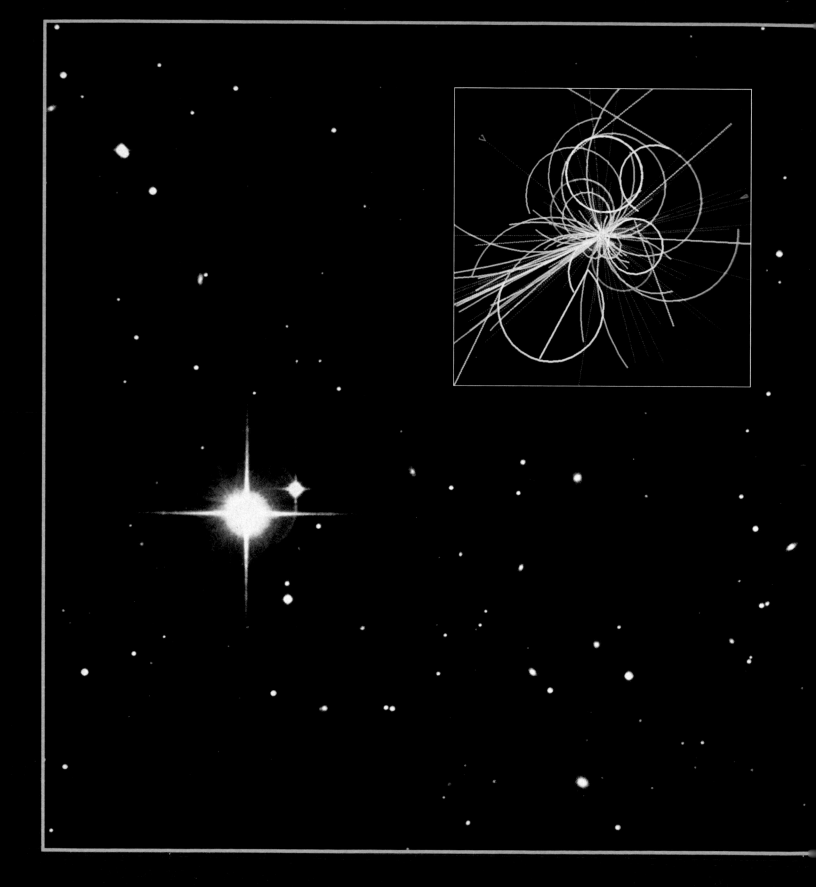

GRAVITY

The quest for a theory that unifies the four fundamental forces of nature must embrace not only the cosmic—represented here by the members of galaxy cluster Klemola 44 *(below, center)*, gravitationally bound to one another across millions of light-years—but also the subatomic, exemplified by the spray of particles produced in a high-energy collision between two protons *(inset)*.

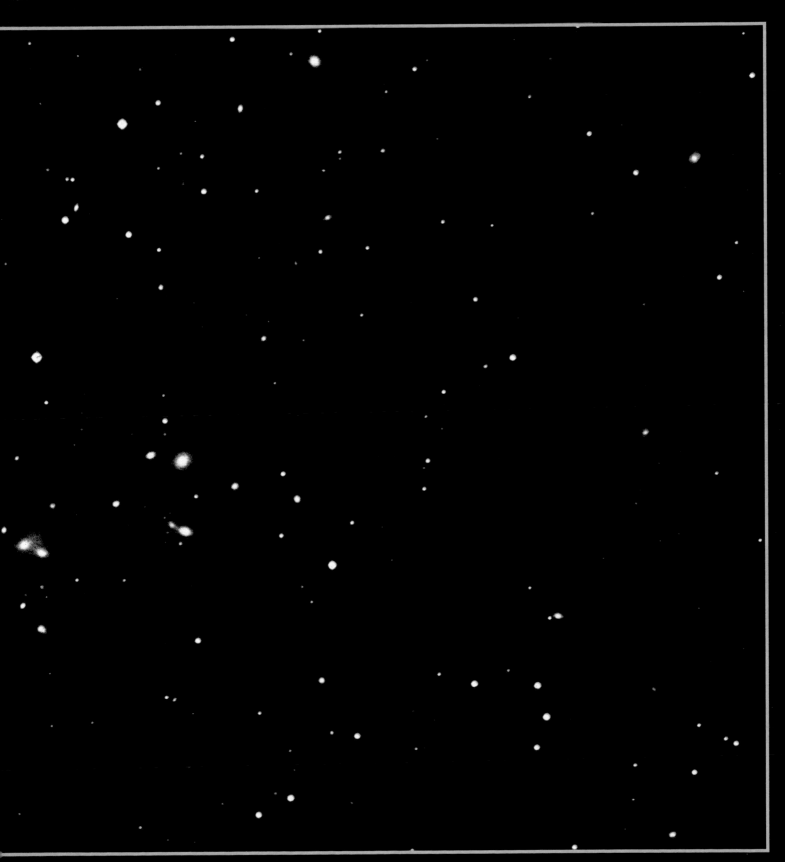

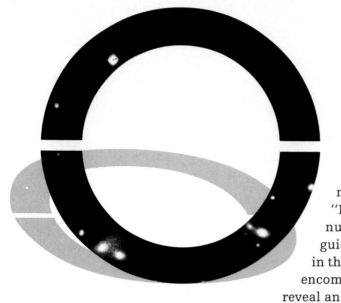

ne piece of popular advice— "Keep your eye on the doughnut, not on the hole"—has guided generations of scientists in their attempts to devise an all-encompassing theory that would reveal an underlying unity behind the physical world's diversity. From Newton to the Grand Unification Theory researchers of the 1970s, physicists have placed primary emphasis on matter and the forces acting on it, and have considered space itself as little more than a blank stage on which the drama of physics plays out. But by the 1980s, some investigators were intent on turning the old adage on its head. The universe could best be understood, they decided, by concentrating not on the "doughnut" of matter, but on the "hole" of apparently empty space. According to these latter-day scientific revolutionaries, just as a doughnut is, in a sense, defined by the presence of a hole in its middle, the existence and behavior of matter in the universe may be intimately connected to the properties of space.

The concept that space is more than mere emptiness, that it can indeed have properties of its own, traces back to the early years of the century and to the achievements of some of the most heralded names in physics. Einstein's special theory of relativity, published in 1905, introduced the notion that space is actually a four-dimensional phenomenon consisting of three spatial dimensions and one of time; to locate an object in this space-time continuum, one must specify not only where it is in relation to some frame of reference, but also when it is there. Einstein spent the next ten years working to extend the special theory to include gravity. The results of his efforts were profound.

Einstein showed that the effects of acceleration are indistinguishable from the effects of gravity, thus shifting relativity's focus to the sphere of gravitational fields and the overall structure of space, which was now beginning to take on an almost tangible quality. The general theory of relativity he eventually completed in 1915 called for a wholesale reinterpretation of Newton's law of universal gravitation in light of the understanding that massive objects literally warp the fabric of space-time. For example, a ball thrown horizontally, which curves toward the ground under the influence of Earth's gravity, should not be thought of as being deflected from a straight-line trajectory through space but rather as following the shortest possible path along space's warped surface.

Space undergoes an even odder transformation when viewed in terms of quantum theory, the other great pillar of modern physics. Max Planck's 1899 discovery that radiation comes in discrete packets established a set value known as the quantum of action, or Planck's constant, a figure equal to the ratio between a photon's energy and its frequency. By combining this constant with the speed of light and a measure of the strength of gravity, one can obtain fundamental units of measurement for both distance and time, which scientists have adopted as absolute standards. The Planck unit of length is on the order of 10^{-33} centimeter, more than a billion trillion times smaller than the diameter of an atomic nucleus. Were it possible to observe space's inherent structure at this incredibly tiny scale, it would take on a surprising appearance. In contrast to the view through relativistic eyes—which shows the space-time fabric as smooth and continuous, like the surface of the ocean seen from a great height—space up close, at the quantum level, theorists speculate, is choppy, billowing and seething like storm-driven waves seen from the deck of a ship.

The central problem of modern physics, of course, is reconciling relativity's cosmic perspective with the subatomic focus of quantum mechanics. Mathematically speaking, relativity provides a perfect description of how gravity curves space, but quantum mechanics has failed to explain what is happening at the Planck scale to account for that curvature. Simply put, a true Theory of Everything, or TOE, must find a way to quantize gravity—to depict the gravitational force in terms of interactions among elementary particles, as quantum physics does for electromagnetism and the strong and weak forces.

Searching for such a comprehensive quantum theory has led researchers to strange and astounding conclusions, including the possibility that there are more than four dimensions and that whole other universes exist. But time and again, the mathematics has fallen short. During the 1980s, however, a promising new theory rose to the forefront of physics, prompting some of its advocates to declare that the elusive TOE was within reach. Known as superstring theory, it proposed that the fundamental constituents of nature are not particles but one-dimensional structures called strings, and that elegant mathematical truths about these structures are reflected in the very fabric of the universe. To some observers, superstring theory represented the most important breakthrough in physics since the days of Einstein and Planck; to others, it simply marked another milestone on the long and difficult road to a satisfactory understanding.

DIMENSIONAL GROUNDWORK

From the moment of its introduction in 1915, Einstein's general theory of relativity created a stir among scientists. Some began to test its formulas and equations, while others spent long hours just trying to work through all its implications. Cosmologists were particularly inspired; the general theory gave them a fresh angle from which to examine gravity's large-scale influence on the shaping and evolution of the universe. However, its approach was intrinsically unlike the method that was employed to study the other known force of the day, electromagnetism. That it took physics off in a totally different direction signaled the fundamental rift that has plagued efforts to unify the forces ever since.

Nevertheless, within a few short years, an imaginative piece of mathematical tinkering hinted that there might be a way to heal the division. In 1919, an obscure mathematician by the name of Theodor Kaluza, who was an unsalaried tutor at the University of Königsberg, began to play around with Einstein's formulas for gravity. For no particular reason other than mathematical curiosity, he reworked the equations to see how they would look in five, rather than four, dimensions. The exercise generated an extra set of equations, which turned out to be the same as James Clerk Maxwell's equations for the electromagnetic field. By assuming the presence of another dimension, Kaluza had produced a mathematical unification of gravity and electromagnetism. But his fifth dimension was purely speculative and had no basis in what was known about the real world. If an extra dimension existed, critics wondered, where was it and why had it never been detected?

Seven years later, Swedish physicist Oskar Klein provided a plausible answer. He explained that Kaluza's fifth dimension might be of such minute range that it was completely imperceptible even at a subatomic scale. He used the example of an extremely thin tube, which from a great distance appears to be a two-dimensional line. What seems to be a point on that line is, in reality, a tiny circle. Thus, any point in three-dimensional space might actually be a circle wrapped around an undetected fourth spatial dimension.

Kaluza and Klein's work would eventually play an important role in unification efforts. Today, in fact, all representations of force fields or the

The first to propose the existence of dimensions beyond the four of Einsteinian space-time was German mathematician Theodor Kaluza *(far left)*, who unified general relativity with electromagnetism in 1919 by incorporating a fifth dimension. In 1926, Swedish physicist Oskar Klein tied the idea to quantum theory, proposing that the fifth dimension was curled up so tightly as to be undetectable in space-time.

architecture of space in more than four dimensions are classified as Kaluza-Klein theories. At the time, however, their insight received scant attention. Physicists of the 1920s were too busy coping with the newly discovered attributes of quantum mechanics, which would have its own strange tales to tell about the nature of space.

SPACE ODDITIES

In 1927, a year after Klein offered his explanation of the possible whereabouts of Kaluza's fifth dimension, Werner Heisenberg spelled out the workings of the quantum world with his postulation of the uncertainty principle, which maintains that it is impossible to measure an atomic particle so that both its location and its momentum can be determined simultaneously with precision. An electron, for example, can appear as a wave—a packet of energy spread out over an indeterminate volume of space—or as a particle, with a definite location but indeterminate momentum. The smaller the region of space being measured, and the shorter the time interval of the measurement, the greater the uncertainty about the electron's behavior.

The central paradox of the uncertainty principle hinges on this problem of scale, and the utter inability of common sense to provide any guidance at the quantum level. Erwin Schrödinger, one of the handful of physicists who helped develop quantum mechanics, illustrated that paradox in a famous thought experiment. A cat is placed in a closed, opaque box, along with a radioactive substance, a Geiger counter, and a vial of cyanide gas rigged to break as soon as the counter registers the decay of a radioactive particle. Because uncertainty rules atomic behavior, quantum physics cannot say whether a given particle will or will not decay within a given amount of time; the outcome can only be stated as a probability. According to the uncertainty principle, until an observation is made, the state of the particle is not just unknown but quite literally undetermined. Thus, Schrödinger's unfortunate cat is neither alive nor dead until someone opens the box to look. In the familiar, everyday world of human experience, a cat must be either alive or dead even if it cannot currently be seen; at the quantum scale, however, it is neither. It would seem, then, that the mathematical truth of quantum mechanics can only be understood by abandoning truth as it is normally perceived by human beings.

Of necessity, such an outlook opened the door on a number of counterintuitive possibilities pertaining not only to the behavior of particles but also to the composition of space. For example, Paul Dirac's 1928 equations describing quantum electrodynamics—a quantum theory of the interactions of electrons and photons *(page 22)*—included solutions that implied the existence of negative energy states. Because a basic tenet of quantum physics is that particles always seek the lowest possible energy level, Dirac's equations seemed to imply that all the particles in the universe should be crowding into every available negative energy state. His solution to this quandary was to propose that all negative energy states are already occupied by hypothetical antiparticles. This prediction was confirmed a few years later when real antiparticles were discovered. Apparently, then, space was teeming with a previously unimagined class of antimatter counterparts to the known particles of matter and energy.

But there were even odder revelations. Dirac's theory of quantum electrodynamics stated that the electromagnetic force operates through the exchange of so-called virtual photons, which pop into and out of existence in an imperceptible flicker of time by means of quantum fluctuations—the central mechanism of the uncertainty principle. This led to the proposition that the creation of something from nothing is a perfectly real possibility: Because the amount of energy contained in a quantum-scale volume of space can never be known with precision, there may be none at all or there may be enough to create a particle of matter. Beyond the level of observation, at the scale of quantum fluctuations, the vacuum of space is filled with all sorts of virtual particles, teamed in matter and antimatter pairs and existing for too short a time ever to be detected. Another way of thinking of these virtual particles is that they represent the probabilities inherent in the uncertainty principle for how real particles will behave.

The implications for cosmology are particularly astounding. One afternoon in 1973, Edward P. Tryon, a physicist at Hunter College in New York, was musing about these quantum fluctuations of space. He realized that, theoretically speaking, quantum fluctuations can be of any size, and that large ones—giving rise to more energetic particles—are merely less frequent than the smaller ones apparently happening all the time throughout space. It suddenly occurred to him that a virtual particle of the largest imaginable energy popping into existence in this way would be indistinguishable from the Big Bang. Excited by the concept, Tryon quickly wrote a paper for *Nature* in which he proposed that the universe may have been born as a fluctuation in "the vacuum of some larger space in which our Universe is imbedded. In answer to the question of why it happened, I offer the modest proposal that our Universe is simply one of those things which happen from time to time."

Tryon's conjecture regarding the birth of the cosmos was certainly intriguing, and it stirred a good deal of additional theoretical exploration by other cosmologists. But it offered little new insight on the precise nature of space itself. Physics still faced the problem of resolving the discrepancy between relativity's treatment of space and the speculations of quantum mechanics.

A NEW LANGUAGE

Traditionally, investigations of how objects move through space and also how matter curves space's fabric have relied on the language of geometry—the branch of mathematics that deals with points, lines, angles, and shapes such as circles, squares, spheres, and cubes. Central to geometry, of course, is the measuring of distances between specific points and yet, paradoxically, space is viewed as consisting of an array of points, such that along a line between any two points there are an infinite number of other points with no meas-

EXAMINING THE HOLE IN THE DOUGHNUT

In their attempt to unify the theories of relativity and quantum mechanics, some physicists are examining topology—the study of the properties of geometric shapes as they are manipulated in various dimensions, focusing on the relationships between surfaces. As illustrated here, for example, a cube is topologically the same as a sphere because both have surfaces with no discontinuities, or holes; either may also be reshaped into a bowl without breaking the surface. A box with no top or bottom *(middle row)*, however, is topologically different from the cube because its surface has been broken. It is thus identical to a doughnut and to a cup with a handle, each of which also has one hole. These in turn, though similar geometrically to the objects in the bottom row, are topologically different because the last row's objects have surfaces with two holes. More complex topologies than these may explain the physical whereabouts of the extra dimensions called for in the mathematics of some unification theories.

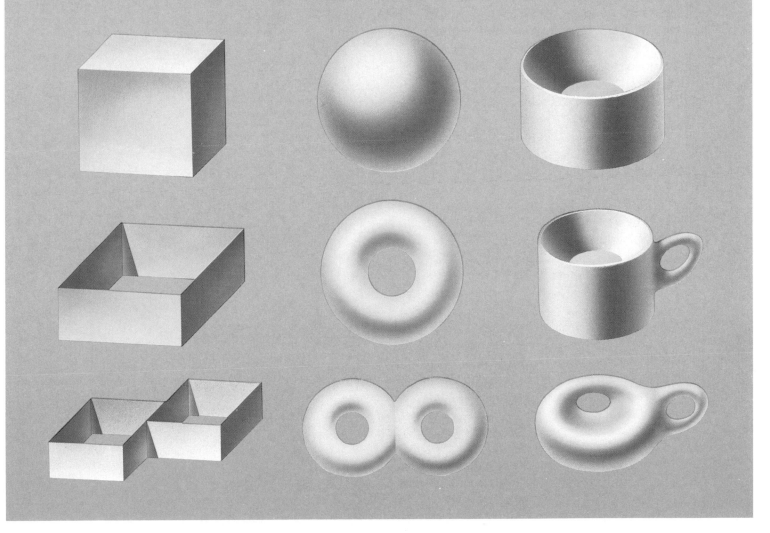

57

urable distance between them. Studies of space thus seem to demand a different mathematical language in which distance is, in essence, irrelevant.

One such language is topology, which began to attract attention in the 1950s as an alternative to geometric techniques. Topology is a mathematical approach that considers more basic properties of objects than their geometric shape. In topology, a square, a triangle, and a circle are identical because one of them can be transformed into another by the appropriate amount of stretching. Rather than concentrating on specific angles and curves, topology focuses on the relationships between surfaces, such as how they might intersect or enclose one another, and whether there are any discontinuities in a given surface—holes, in other words. Topologically speaking, a sphere and a doughnut are fundamentally different because one has a hole and the other does not; transforming a sphere into a doughnut would require not only stretching but also tearing the sphere.

By ignoring the detailed geometric description of a surface's curves and angles, topology looks at space according to a different set of parameters. As a result, it has opened up new avenues for the investigation of space, leading researchers to some of the most bizarre conclusions in the whole history of theoretical physics. Complex topologies, for instance, allow extremely intricate spatial surfaces to be enfolded within a tiny region; within the folds of such a space, distant regions of the universe might actually come in contact with one another. This notion prompted Princeton physicist John Archibald Wheeler to introduce the concept of the wormhole in 1957. A rip or tear in the fabric of space—equivalent to the topological transformation of a sphere into a doughnut—would open up a hole connecting two widely separated locations. Such holes could be produced, said Wheeler, by quantum fluctuations in the topology of space at the Planck scale.

However, the connection between a topological outlook and the unification

Japanese physicist Yoichiro Nambu offered the first string theory in 1970. He proposed that elementary particles were actually tiny strings that vibrated at certain energies, thereby accounting for a variety of particles, much as violin strings vibrating at different rates produce a variety of musical notes. However, the theory required twenty-six dimensions.

of gravity with quantum mechanics remained unclear into the 1960s, and mainstream physicists concerned with achieving a single, unified view of reality continued to concentrate on more traditional approaches. But a few scientists and mathematicians kept tinkering with concepts involving topology and dimensionality. Their angle of attack was inherently different. Most physics theories grow from the bottom up, building on a foundation of observed facts and drawing general conclusions from them. The maverick hypotheses evolved from the top down, starting from a framework of other mathematical theories and only later considering whether reality might confirm their assumptions.

The top-down approach, useful though it is in creating new insights, has produced a philosophical schism within the physics community, pitting experiment-oriented physicists against those who search for truth in the realm of pure mathematics. This conflict has never been resolved to anyone's complete satisfaction, but even die-hard experimentalists have acknowledged the mathematical elegance of the theorists' work.

ON THE ROAD TO STRINGS

The first steps toward a radically new way of looking at the world took place in 1968. Italian physicist Gabriele Veneziano (now at CERN, the European Center for Nuclear Research, near Geneva) and a colleague were thumbing through some old mathematics texts when they stumbled on a formula by nineteenth-century Swiss mathematician Leonhard Euler. Veneziano was amazed to realize that this formula seemed to be the solution to a contemporary problem involving the interactions of hadrons, the family of elementary particles that are subject to the influence of the strong nuclear force. Unable to describe the precise details of such particle interactions, physicists simply accepted that the interactions themselves took place, in effect, inside a black box. They were more interested for the moment in tracking the results of interactions and collisions between particles than in understanding the actual mechanics; in other words, all that mattered was what went into the box and what came out of it.

To plot these outcomes, physicists had been employing an organizational system known as the scattering matrix, or S-matrix (so called because the particles that result from high-energy collisions are made to scatter in various directions). Veneziano found that he could use Euler's formula to create a mathematical model that described certain important relationships in the S-matrix and accurately predicted the results of particle interactions. Through this model, the S-matrix became a much more valuable tool for understanding the workings of the strong force and offered the possibility of revealing what was going on inside the black box of actual interactions.

But Veneziano's model had a disturbing flaw: It allowed for the existence of what physicists whimsically call ghosts—outcomes with nonsensical negative probabilities. (It is meaningless, for example, to say that there is a −40 percent chance of rain.) The ghosts could be exorcised only by reformulating the equations in a mathematical space containing twenty-six dimensions.

To a mathematician, a dimension is simply a set of coordinates in space, used to specify a location. The position of an airplane in the ordinary four-dimensional world of space-time, for example, can be specified with four sets of coordinates—one each for longitude, latitude, altitude, and time. In some forms of mathematics, such as topology, additional coordinates, or dimensions, can be added to provide even greater specificity about location. By using twenty-six sets of coordinates to define positions within the S-matrix, Veneziano was able to eliminate the troubling ghosts.

This positing of twenty-six dimensions was a classic example of top-down reasoning, since it was a purely mathematical solution with no analogy in the real world of actual particles. Nevertheless, Veneziano's model did provide inspiration, primarily because it turned out to be similar to other formulas describing how objects vibrate. Violin strings, for instance, vibrate in a particular way to produce musical notes. The specific notes produced depend on the tension in the string; in effect, the string stores energy and releases it at certain discrete intervals. In quantum terms, the notes can be thought of as unique localizations of energy—in other words, as elementary particles. In 1970, Yoichiro Nambu, a brilliant Japanese physicist who was then at the University of Chicago, noticed this correspondence between the mathematics of vibrating strings and Veneziano's equations for the behavior of hadrons. He realized that Veneziano's model was the equivalent of describing hadrons as if they actually were strings, rather than particles. Nambu proposed that such strings were real, physical entities, with a length of 10^{-13} centimeter (about the size of a proton), and that their vibrations were responsible for what had been observed as the various types of hadrons.

Nambu's strings, aside from being invisibly small, were also massless and elastic, and whipped around at close to the speed of light. Their vibrations and rotations followed the same quantum rules that dictate the discrete spin values of particles. According to Nambu's theory, differences in these characteristics and in a string's energy level accounted for different kinds of hadrons—some of which are fermions, or matter particles, and some bosons, or force carriers. A nonvibrating, nonrotating string, with little energy, corresponded to low-mass hadrons such as pions—the bosons that mediate the strong force in atomic nuclei, holding protons and neutrons together. Vibrating, rotating strings at higher energies corresponded to hadrons with greater mass, including such fermions as the protons and neutrons themselves. All of this was in agreement with the predictions of Veneziano's hadron model; even better, it was also in agreement with the types of hadrons that had actually been observed in accelerator experiments.

STRINGS SIDETRACKED

Suddenly, string theory became the hottest thing in physics. However, it suffered from a great many loose ends, both literally and figuratively. For one thing, there was the question of how strings related to quarks. In 1964,

Flatland. One of the problems inherent in theories that posit more than three spatial dimensions is that the extra dimensions are as difficult to visualize in a three-dimensional world as would be a three-dimensional object in a flat, two-dimensional world (right). A cone passing through a two-dimensional plane would be perceived by observers in that plane first as a point, then as a parabola, and later as a triangle. Thus, in two-dimensional space, the third dimension would be comprehensible only in mathematical terms—precisely the situation confronting physicists whose theories call for ten or more dimensions.

Murray Gell-Mann and George Zweig had independently demonstrated that hadrons are made up of combinations of classic pointlike particles called quarks, yet string theory said that hadrons were extended, one-dimensional vibrating objects. Nambu and his colleagues initially suggested that perhaps strings were themselves the quarks, but that did not make sense because strings were about the same size as protons and therefore obviously could not fit inside protons, which are composed of three quarks. As an alternative, Nambu proposed that perhaps quarks sat at the ends of strings. That would explain why quarks had never been observed in isolation; a free quark would be the equivalent of a string with only one end. Then again, three-quark hadrons such as protons and neutrons would require a three-ended string. That might be achieved by splitting a string into a Y shape, but it was not a very elegant solution.

A more serious problem with Nambu's string theory was that it did not in fact encompass all hadrons. Both fermions and bosons possess the quantum property known as spin, but for bosons the spin value is always an integer, such as 0, 1, or 2; for fermions, the value of the spin is always a fraction, such as 1/2 or 3/2. Nambu's theory, it turned out, could only account for integer spin and thus applied only to those hadrons that are bosons. In early 1971, Pierre Ramond, then at Yale and now at the University of Florida, developed a way of formulating the string equations so that they would accommodate fermions. Meanwhile, John Schwarz at Princeton University and André Neveu in France were tackling the same problem and came up with a solution that incorporated the Ramond calculations. The result was an all-embracing string theory that included every type of hadron. As a bonus, the Ramond-Neveu-Schwarz theory required only ten dimensions (nine of space and one of time), instead of the cumbersome twenty-six that Nambu had inherited from Gabriele Veneziano. This was clearly an important

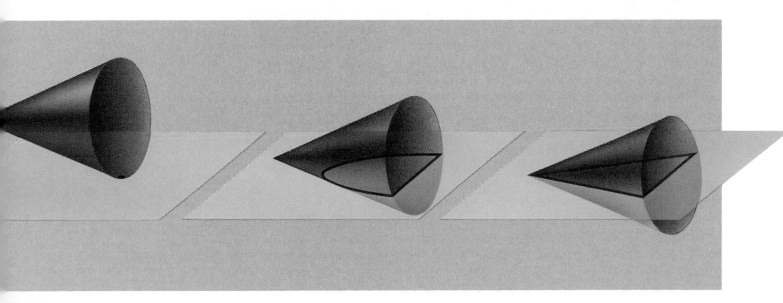

step in the right direction, toward the familiar world of four dimensions.

Even as string theorists were making progress in elucidating their concept, they were losing support in the broader community of physicists. Strings operating in ten dimensions of space-time were still just too complex for most physicists to accept. By 1973, as the initial enthusiasm for strings was fading, the theory of quantum chromodynamics—which satisfied many lingering questions about the workings of the strong force—took center stage *(page 34)*. Quantum chromodynamics (QCD) did a much better job of describing strong-force interactions, and its straightforward treatment of quarks as traditional point particles seemed more plausible than a model involving quarks whipping around at the ends of strings. "String theory," John Schwarz later recalled, "dried up practically overnight."

KEEPERS OF THE FAITH

In 1974, with string theory apparently dying, Schwarz—who had moved to the California Institute of Technology in 1972—arranged a meeting with French physicist Joel Scherk, with whom he had worked previously at Princeton. Although most strong-force investigators had switched their attention to QCD, Schwarz and Scherk had staked an early claim to strings, and they were reluctant to give them up. "We felt strongly that string theory was too beautiful a mathematical structure to be completely irrelevant to nature," Schwarz later wrote. "We still thought that string theory deserved a last look before being abandoned."

Among the presumed defects of string theory was that it posited the existence of all sorts of particles that had never been observed. But one type of these particles had enticing characteristics. According to the string equations, its properties precisely matched those of the hypothetical quantum of gravitation, the graviton, thought to be responsible for the curvature of space. As a particle physicist, Schwarz had paid little heed to gravity, whose force is considered to be negligible at the level of elementary particles. Nevertheless, he and Scherk gradually came to realize that the presence of the graviton in string theory might be its savior, if the theory could be turned to a purpose different from its original one, which was to explain the characteristics and behavior of strong-force particles.

Schwarz and Scherk proposed that string theory might be a means for unifying gravity and the other fundamental forces. In order to get gravity from strings, all that had to be done was to shrink the strings by twenty orders of magnitude, from the 10^{-13} centimeter of Nambu's original conception to the 10^{-33} centimeter of the Planck scale. The two physicists believed they were onto something very important, and throughout 1974 and 1975, they eagerly wrote papers and gave lectures to spread the word. The physics community greeted their efforts with a yawn. "For the most part," wrote Schwarz, "our work was politely received—as far as I know, no one accused us of being crackpots. Yet, for a decade, almost none of the experts took the proposal seriously."

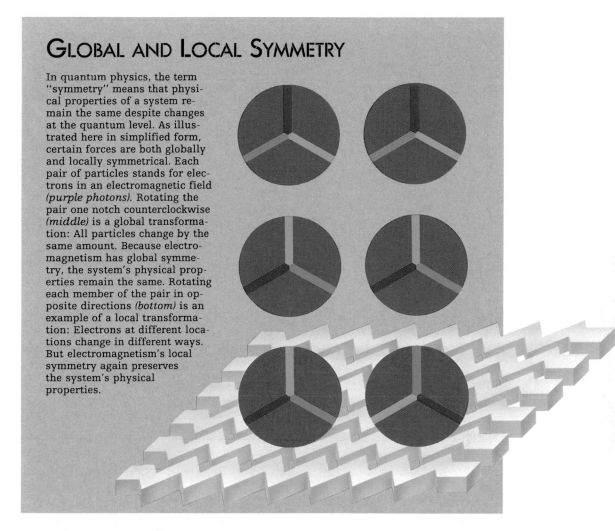

GLOBAL AND LOCAL SYMMETRY

In quantum physics, the term "symmetry" means that physical properties of a system remain the same despite changes at the quantum level. As illustrated here in simplified form, certain forces are both globally and locally symmetrical. Each pair of particles stands for electrons in an electromagnetic field *(purple photons)*. Rotating the pair one notch counterclockwise *(middle)* is a global transformation: All particles change by the same amount. Because electromagnetism has global symmetry, the system's physical properties remain the same. Rotating each member of the pair in opposite directions *(bottom)* is an example of a local transformation: Electrons at different locations change in different ways. But electromagnetism's local symmetry again preserves the system's physical properties.

THEORY AND SUPERTHEORY

Although string theory seemed to be falling by the wayside, a host of other new theories were coming into vogue and offering novel approaches to the most basic problems of physics. Perhaps the most important of these was the idea of supersymmetry—an elaboration on the concept of symmetry that had proved so instrumental in the development of QCD and in successful efforts to unify the three quantum forces.

There are two types of symmetry, global and local, both of which concern the preservation of a given attribute or characteristic of a system despite a change in some other characteristic. Global symmetry is fairly unsophisticated because it involves identical changes everywhere throughout a system, like the reversal of an image in a mirror. Local symmetry, however, means that the underlying relatedness between diverse phenomena is maintained despite localized, or different types of, changes across the system. Local symmetry, which when applied to force fields is known as gauge invariance, is more significant to physics because it can explain how dissimilar forces can be unified despite their observed differences. The electromagnetic, strong, and weak forces are called gauge forces because they all exhibit local symmetry, and this is what made it possible for researchers in the 1970s to synthesize them in the Grand Unification Theories. But gravity eluded unification until a new kind of symmetry—supersymmetry—was applied.

Among several physicists who independently came up with the concept of supersymmetry in the early 1970s were Julius Wess, of Karlsruhe University in Germany, and Bruno Zumino, of CERN. The two had once worked together at New York University, but inspiration did not strike until they attended a seminar on strings at CERN in 1973. After listening to a talk about symmetrical relationships inherent in strings, the pair began to wonder if such symmetries could apply to a wider class of particles and not just the hadrons encompassed by string theory. The Ramond-Neveu-Schwarz theory had shown that strings could produce both half-spin fermions and integer-spin bosons. Wess and Zumino found a way to bring about a mathematical unification of all types of fermion and boson particles without reference to strings. To achieve this, they applied a wholly new concept of space called superspace, developed by Abdus Salam and John Strathdee of the International Center for Theoretical Physics in Trieste, Italy, which tacked on four new dimensions to the familiar four of space-time. Viewed in this eight-dimensional context, fermions and bosons were indistinguishable from one another. The concept dramatically simplified the universe by showing that the two seemingly separate classes of particles were really part of the same family, linked by supersymmetrical relationships—that is, superspace's counterpart of ordinary symmetries.

With the creation of the "supersymmetry" theory in 1973, Bruno Zumino *(near right)* and Julius Wess discovered that they could mathematically change bosons (force particles) into fermions (matter particles) and vice versa, thereby reducing fundamental particles from two families to one—a long step toward unifying all four fundamental forces. The theory posited eight dimensions and the existence of superparticles, including the gravitino, superpartner of the gravity-carrying graviton.

Naturally enough, the elegant mathematical logic was not without its complications. Supersymmetry relies on the prediction that every fermion and every boson possesses a hypothetical supersymmetrical partner, particles that came to be identified by the addition of an "ino" suffix or an "s" prefix. Thus, photons were linked with photinos, gluons with gluinos, quarks with squarks, electrons with selectrons, and so on. The apparent absence of such particles suggested that in the real world, supersymmetry had somehow been broken. The most plausible explanation was that the unbroken symmetry of superparticles and their partners existed only in the excessively hot universe of the Big Bang era, and is broken in the cooler, lower-energy state of the universe as it exists today. To restore supersymmetry and find the superparticles, particle accelerators would have to be able to reproduce energies, or temperatures, close to those that existed in the first instants following the Big Bang. The theory thus remained untested.

Meanwhile, a handful of researchers had been examining the question of whether supersymmetry exhibited local symmetry. If it did, they reasoned, then it would have an entirely new gauge field and a new force-carrying particle associated with it; calculations indicated that these would correspond to the gravitational field and its hypothesized quantum particle, the graviton. To physicists, this was an exciting development. Armed with a new superpartner particle for the graviton—labeled the gravitino—and a new theory of so-called supergravity, some physicists believed that they were closing in on the holy grail itself, a unified field theory that included all four fundamental forces of nature.

But the supergravity approach was not unique. Several competing supersymmetry models—based on different arrangements, or symmetry groups, of particles and superparticles—were beginning to emerge, some implying the existence of eleven dimensions. One of the theories seemed to provide a solution to the troubling notion of all those additional, unobserved dimensions. It described a process that, early in the universe's history, caused the extra dimensions to compactify, in physicists' jargon, or curl up into a minute structure and effectively disappear from view. Suddenly, the venerable Kaluza-Klein theory of nearly fifty years earlier was taking on a new relevance. Multidimensionality and the inability to perceive it might be merely a matter of a limited perspective inherent in the present state of the universe.

Supersymmetry was not without relevance for the overlooked theory of strings. In 1976, Joel Scherk, along with Italian physicist Ferdinand Gliozzi and David Olive of the Imperial College of Science and Technology in London, published a paper that incorporated the tenets of supersymmetry, reworking string theory so that it would accommodate the existence of the gravitino, in addition to the graviton. In effect, their version transformed the Ramond-Neveu-Schwarz strings into superstrings. The development was interesting to the few remaining string die-hards, but to practically no one else. By the late 1970s, most researchers were in hot pursuit of the Grand Unification Theories and cared little about supersymmetrical refinements in a theory that was considered old hat. To the vast majority of physicists, strings—even reconfigured as superstrings—were still yesterday's news.

THE EMERGENCE OF SUPERSTRINGS

John Schwarz, however, retained his belief in the beauty and the potential unifying power of string theory. He was not alone—not quite. Michael Green, of Queen Mary College, University of London, had done his Ph.D. dissertation on Veneziano's theories in the late 1960s. Like Schwarz, he had never abandoned an early romance with strings. According to Green, "String theory is, once you have learnt it, so captivating, so elegant, that it's very difficult to put it out of your mind." Then, in the summer of 1979, the two men happened to be visiting CERN at the same time. Over a cup of coffee, they discussed their shared fascination and agreed that the theory was worthy of further exploration. Specifically, they decided to concentrate on the supersymmetry-string connection that had been proposed three years earlier by Scherk, Gliozzi, and Olive. Their efforts that summer produced little besides a resolve to continue their collaboration.

In 1980, Schwarz and Green met again at the Aspen Center for Physics, a kind of summer camp for theoretical physicists. Freed from the distractions of teaching duties and invigorated by the mountain air, they once more tackled the problem of superstrings. This time, they managed to resolve certain mathematical anomalies in their supersymmetry model, and by the end of their stay had come up with a coherent theory describing the properties of superstrings existing in a ten-dimensional version of space-time. They labeled this the Type I theory, and followed it up the next summer with the Type II theory, which described the special case of superstrings whose ends

Spurred by the work of brilliant Princeton physicist Edward Witten *(right)*, superstring theorists John Schwarz *(far left)* of the California Institute of Technology and Michael Green *(left)* of Queen Mary College, London, found a way to incorporate gravity into string theory in 1984 without troublesome mathematical anomalies. Witten's prompt and enthusiastic endorsement of their work gave superstring theory long-sought credibility in the physics community.

were joined, forming closed loops. To the few physicists who were still paying attention, the marriage of supersymmetry and strings seemed natural and obvious. "Everybody was making everything supersymmetric in those days," Green later recalled, noting that he and Schwarz were "mesmerized" by the possibilities that seemed to be inherent in superstrings. As their work progressed, they became more and more certain that their investigations would eventually pay off. The work was exhilarating. As Green put it, "I never worked with such intensity in my life."

The cardinal defect in other theories attempting to apply quantum physics to gravity had been the appearance of infinite values in the solutions. But Schwarz and Green found that replacing dimensionless point particles with one-dimensional strings magically did away with the problem. Type I superstring theory could be renormalized—its infinities mathematically canceled out—just as quantum field theories for electromagnetism and the strong and weak forces had been. Even more significant, the Type II theory yielded only finite results. Schwarz and Green announced their findings in a series of papers, hoping that other physicists would finally take note. In a 1982 article in *Physics Reports*, Schwarz virtually begged for recognition: "The limiting factor has been the time available to Green and myself in the last two years. Hopefully, this article will encourage others to become involved."

It did not. Green was not all that surprised. "Most people just simply felt string theory was too far away from conventional quantum field theories," he recalled. The stumbling block seemed to be the necessity of explaining how ten-dimensional superstrings had been compactified into the four dimensions of recognizable space-time. The mathematics of compactification were admittedly complex, having to take into account all the forces and particles

of the four-dimensional universe, not just gravity. In fact, Green later admitted, superstring theory "didn't look as though it could possibly make contact with the laws of physics, apart from the force of gravity." Given these shortcomings, superstring theory failed to win many converts. But help was on the way.

Despite the lack of popularity of Schwarz and Green's work, a few prominent physicists kept up with it, among them, Edward Witten, universally regarded as one of the most perceptive of today's theorists. His brilliance has illuminated many corners of science, including cosmology and particle physics. A history major as an undergraduate at Brandeis, Witten flirted with a career in journalism and politics before deciding that he lacked the "common sense" presumed to be necessary in those fields. Instead, he entered graduate school at Harvard as a physics student. By the age of 28, in 1979, he had become a full professor at Princeton. In 1982, with his colleague Luis Alvarez-Gaumé at Harvard, Witten completed an important paper that tackled some of the most difficult concepts in theoretical physics. At first, the paper seemed only to add to the problems confronting Schwarz and Green, but its reasoning ultimately inspired a solution.

BREAKTHROUGH

Witten and Alvarez-Gaumé's research addressed, among other things, an issue that had first emerged in the late 1950s, when experimenters discovered that the weak force sometimes differentiates between particles that spin in a right-handed direction and those that spin in a left-handed direction *(pages 40-43)*. Originally, physicists had assumed that there was a fundamental symmetry between left-handed and right-handed spin and that the universe was, in effect, even-handed. But this symmetry had clearly been broken. As a result, any theory that incorporated the characteristics of the weak force— and thus all those that sought to unify the forces—would have to include terms accounting for handedness, or chirality (from the Greek word for hand). In the case of multidimensional theories, the correct chirality of particles would have to be preserved through the process of compactification down to the present four observable dimensions.

Before his collaboration with Alvarez-Gaumé, Witten had shown that this could be somewhat problematic. Calculations he performed indicated that maintenance of chirality is possible only in the compactification of an even number of dimensions. But he had also demonstrated that, within the context of Kaluza-Klein theories, the present four forces could arise only through the compactification of an odd number of dimensions. As if that were not enough, the paper Witten coauthored with Alvarez-Gaumé went on to point out that, for complex reasons, most ten-dimensional theories were bound to contain certain anomalies that would lead to unacceptable violations of basic conservation laws—such as the conservation of charge.

For Schwarz and Green, the news was hardly reassuring. Superstring theory, it seemed, had fallen victim to a kind of multidimensional Catch-22. On the one hand, Schwarz and Green's ten-dimensional model preserved chirality, whereas an eleven-dimensional version would not; on the other hand, compactification of eleven dimensions would give rise to the four forces and avoid anomalies—tricks that ten dimensions apparently could not muster. Still committed to superstrings, the two theorists decided to stick with ten dimensions and attempt to root out the anomalies; furthermore, they would try to adapt their model so that it would start out with the gravitational and gauge forces already present in the precompactified ten dimensions of the superstrings. Deciding that their Type II closed superstrings could not meet these criteria, they turned once again to the open, Type I superstrings.

Upon returning to Aspen during the summer of 1984, Schwarz and Green began examining several different versions of their Type I superstring theory, each based on a different symmetry group, or collection of mathematically related particles. Just before one of the seminar sessions, Schwarz remarked to Green that maybe there was a particular symmetry group in which certain key anomalies canceled out. "This expectation," Green later admitted, "was not based on anything more than wishful thinking." Although most of their colleagues seemed to think that string theory would always have the anomaly problem, he and Schwarz were more optimistic. String theory, they thought, "was so magical that it would always get around the anomaly problem." The truth lay somewhere in between. "It turns out that almost all string theories are indeed sick, they almost all possess anomalies," Green noted. "But amongst the theories that we were looking at was a single unique theory which avoided this problem." After the seminar, Green turned to Schwarz and said, "SO(32)." "Then," recalled Schwarz, "we started getting excited."

The two set about reformulating the Type I theory to conform with the requirements of the SO(32) symmetry group, which was one of the largest, containing 496 gauge particles. One by one, anomalies fell by the wayside. Finally, a single calculation remained, a simple multiplication of two key figures. If the answer was 496, their superstring theory would be proven anomaly-free. While Schwarz sat at his desk, Green stood at a blackboard and performed the calculation, multiplying 31 by 16. He came up with—486. "Oh dear," said the Londoner, "it doesn't work." "Try it again," said Schwarz. Green did, and this time successfully completed the third-grade problem. The answer was, indeed, 496.

The new superstring theory was nothing less than what Green said it was: magical. It was supersymmetrical and chiral, it preserved the fundamental forces of the four-dimensional universe, and it had no anomalies. Best of all, it predicted the existence of gravity, something no other candidate for a Theory of Everything had done.

A few days after the session at the blackboard, Schwarz and Green announced their stunning results in an unusual way—at a cabaret show that was part of the conference's entertainment. The method was not without precedent. During a similar diversion at the Center for Physics at Aspen in 1976, quark theorist Murray Gell-Mann had leaped up from the audience and begun talking wildly about how he had just come up with a theory that included quarks, gravity, and everything else in the universe. He went on for several minutes until, on cue, two men in white coats came and dragged the seemingly demented Nobelist away. Now it was Schwarz's turn. "I started ranting and raving about how string theory was going to be the complete theory of nature," he recalled. "And nobody knew anything about that yet, so they thought it was all a joke. Then the guys in the white coats came out and carried me off." On that note of feigned madness, the world of physics entered the age of superstrings.

THE SUPERSTRING REVOLUTION
After the conference in Aspen, Schwarz and Green hurriedly prepared a paper formally announcing their breakthrough. Even before prepublication copies

REDEFINING SPACE

Twistor theory, the creation of Oxford mathematician Roger Penrose, is a radical attempt to synthesize the subatomic world of quantum theory and the general theory of relativity, which describes gravity as the warping of space-time by the presence of mass. Such a synthesis would allow gravity to be unified with electromagnetism and the strong and weak nuclear forces, all of which have been successfully described in quantum terms.

The incompatibility between general relativity and quantum theory is illustrated below with a device known as a light cone, which represents a kind of snapshot of events in space-time. Since light radiates from a source like ripples on a pond, the outward expansion of the circle of light in a given period of time forms a cone, made up of the paths of individual photons. Any point lying on or within the cone is within the sphere of influence of the emitting object—that is, the source and the point may be said to be causally related, and the path between them is part of the so-called causal structure of space.

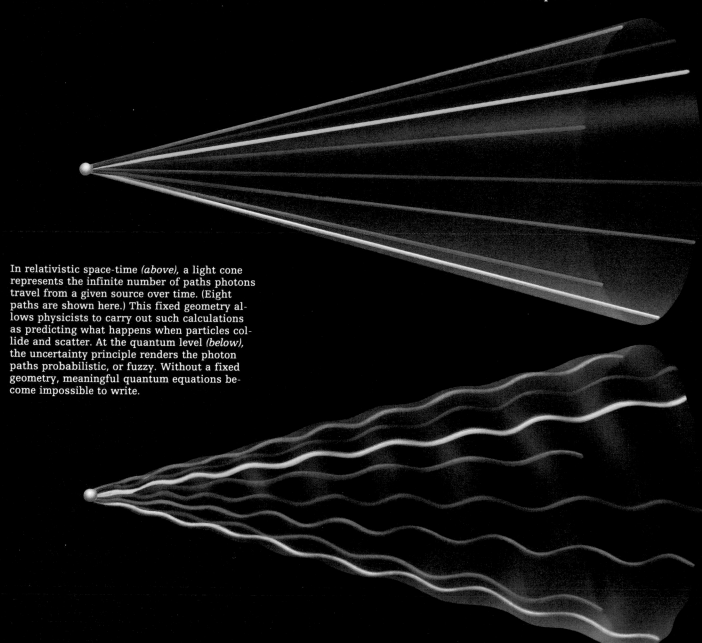

In relativistic space-time *(above)*, a light cone represents the infinite number of paths photons travel from a given source over time. (Eight paths are shown here.) This fixed geometry allows physicists to carry out such calculations as predicting what happens when particles collide and scatter. At the quantum level *(below)*, the uncertainty principle renders the photon paths probabilistic, or fuzzy. Without a fixed geometry, meaningful quantum equations become impossible to write.

In relativity theory, the cone describes the geometry of space-time. The photons travel at the speed of light along paths whose contours are determined by the presence or absence of matter. At the quantum level, however, uncertainty sets in. Because the behavior of subatomic particles cannot be measured with precision, their paths may be expressed only as probabilities, and any causal relationships are equally uncertain, making it impossible to apply general relativity to quantum mechanics.

As illustrated here and on the following pages, twistor theory circumvents some of these difficulties by using a new description of the structure of space itself. Penrose achieved this mathematically by incorporating two features integral to quantum physics and applying them to relativistic space: complex numbers, a mixture of familiar real numbers and the imaginary numbers invented by mathematicians to represent the square roots of negative numbers; and particle spin, the intrinsic rotation of a particle. The result is a complex space with eight dimensions, whose building blocks, twistors, can combine to form both the particles traveling through space and the points that define space.

A twistor, the fundamental construct of twistor theory, is the path traveled in space-time (depicted as a cube at right) by a massless particle, such as a photon. Such a path *(red line)* is called a null line in conventional geometry. In the complex dimensions of twistor space, represented for simplicity by the sphere at far right, the path and all the points along that path are described by one twistor—shown here as a point.

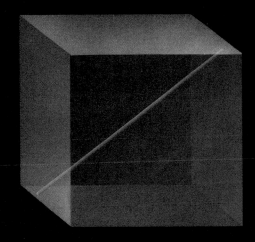 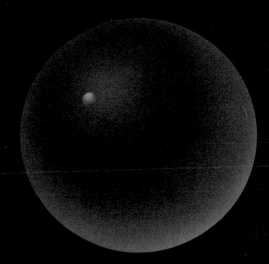

In space-time, a point is defined by the intersection of the set of null lines that would make up the light cone that could emanate from the point *(right)*. Since each null line is a point in twistor space, such an intersection of null lines translates as a series of twistors along a line.

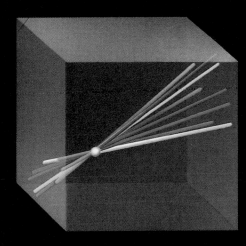 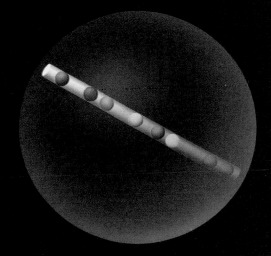

THE POWER OF TWISTORS

Twistor theory's significant breakthrough, illustrated in simplified fashion at right, was finding a way to describe the behavior of massless particles in quantum terms. In the workaday world, the theory allows physicists to translate highly complex differential equations into less complex algebraic equations when solving problems involving the interactions of fundamental particles.

In addition to possessing this useful faculty, twistors may also be the building blocks of the particles themselves. According to Penrose and his co-workers, a single twistor can mathematically represent a massless particle such as a photon. Particles with mass are produced by combinations of twistors: Two can interact to make an electron, three can yield protons and neutrons, and higher combinations generate the heavier entities that are detected fleetingly in particle accelerators.

Ultimately, Penrose hopes that twistors will be able to account for the four fundamental forces. Just as an object with mass warps Einsteinian space-time to produce a gravitational field, a particle could deform twistor space in a way that generates not only gravity but the other three forces as well—electromagnetism and the strong and weak nuclear forces. Such a development would not only profoundly reshape the way physicists think of space, time, and elementary particles but also would mean that science had finally found the Theory of Everything.

The geometry of a light cone in space-time *(right)* is fixed, consisting of definite paths, or null lines, traveled by photons emanating from a definite given point.

When such a light cone is represented in twistor space, the situation is analogous to that of representing a point. Because all of the null lines in the space-time cone intersect at a given point, that point translates into a line that connects pointlike twistors corresponding to each of the null lines.

The advantage of twistor theory becomes apparent when quantum mechanics is brought to bear on the twistor version of the light cone. Because the uncertainty principle does not affect individual points in space but only the relationships between them, the points remain fixed, and the relationships become probabilistic, or fuzzy.

When the quantized twistor light cone is translated back into space-time, the result is quite different from the cone produced when a space-time light cone is quantized *(page 72)*. Because the points in the twistor light cone are unaffected by the uncertainty principle, the null lines to which they correspond in space-time are well defined. But the fuzziness between the points in twistor space manifests itself in the alignment of the null lines, which no longer converge precisely. The misalignment leaves the apex of the light cone fuzzy—a problem that twistor theorists have yet to solve.

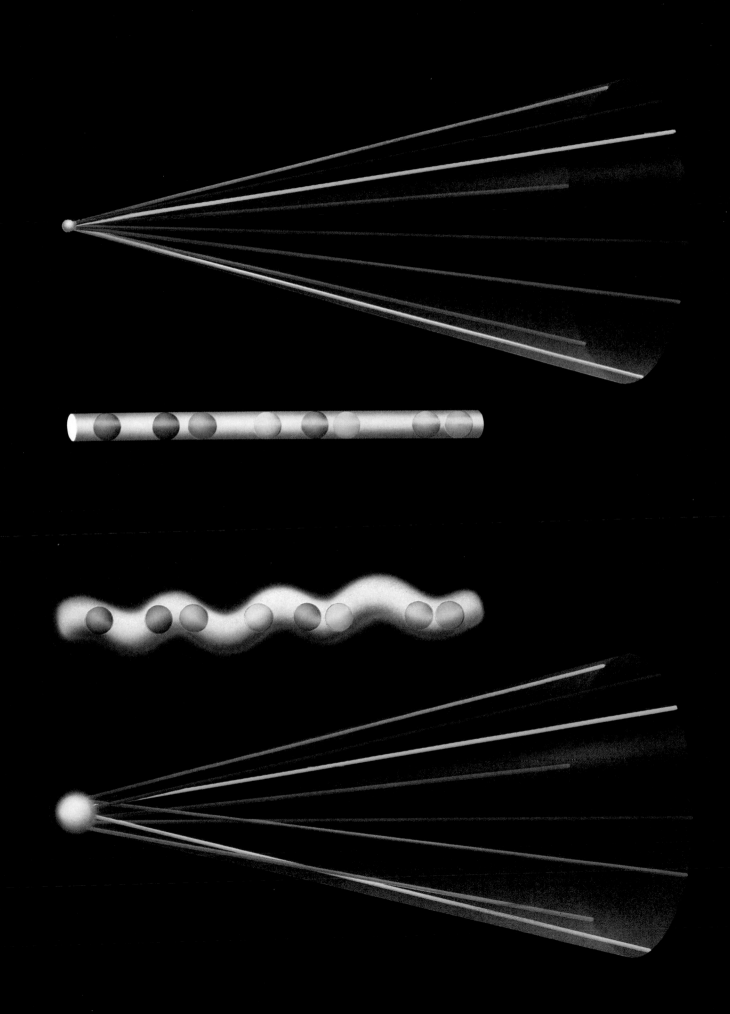

With his advocacy of twistor theory, British theoretical mathematician Roger Penrose of Oxford University is redefining space itself. Twistors, like the one depicted on the blackboard, make up both the points that define space and the particles that inhabit it. The result is a complex, eight-dimensional space that may open the way to a quantum theory of gravity.

half century." Witten's enthusiasm, no less than Schwarz and Green's groundbreaking insight, helped propel this fundamental reworking of basic principles to the forefront of theoretical physics. Superstrings have offered plausible, if not quite ironclad, solutions to many problems that stymied more traditional approaches for decades. Their mathematical elegance and "miraculous" properties exert a powerful appeal and, in a sense, outshine the brilliant contributions of the handful of researchers whose dedication has been unwavering. "I don't think that any physicist would have been clever enough to have invented string theory on purpose," said Witten in the interview. And no one, not even its chief proponents, can admit to fully comprehending its conceptual basis. "We are in a peculiar position with string theory," says John Schwarz, "in that we know some of the equations but we don't really have a deep understanding of the principles that underlie those equations. The history here is backwards."

As a result, more than a few mainstream physicists have been reluctant to jump on the superstring bandwagon. In an interview conducted not long before his death in 1988, Nobelist Richard Feynman expressed his skepticism about superstrings. "I have noticed," he said, "when I was younger, that lots of old men in the field couldn't understand new ideas very well, and resisted them with one method or another, and that they were very foolish in saying these ideas were wrong. . . . I'm an old man now, and these are new ideas, and they look crazy to me, and they look like they're on the wrong track. . . . I would be very foolish to say this is nonsense. I am going to be very foolish, because I do feel strongly that this is nonsense! . . . So perhaps I could entertain future historians by saying I think all this superstring stuff is crazy and is in the wrong direction." Sobering words from a physicist noted for his imagination and open-mindedness.

Indeed, one well-known figure in modern physics, British mathematician Roger Penrose, has worked diligently since the 1960s developing a competing theory that may someday succeed where superstrings still fall short. His approach, known as twistor theory *(pages 72-75)*, has perplexed more than a few in the scientific community but has already survived longer than first seemed likely and in some ways has run into fewer problems than its better-known rival. Other possibilities are also being pursued by some of the most imaginative thinkers in the physics community. Their speculations typically build on the supposition that the very structure of space-time is discrete, that time and causality proceed not continuously but in step-by-step fashion—somewhat analogous to the workings of a computer. Viewed this way, the distinction between the geometry of space-time and the particles that move in it completely evaporates. Although they continue to struggle with establishing the mathematical underpinnings that physics requires, the advocates of this approach feel it holds the greatest promise of uniting the principles of quantum mechanics with those of general relativity and achieving the long-sought synthesis.

A CONTINUING DEBATE

Meanwhile, superstring theory has continued to elicit criticism. One of its fiercest detractors has been Harvard's Sheldon Glashow, who won the Nobel prize for his work on the theory of electroweak unification. In a 1986 article with colleague Paul Ginsparg, he blasted the current fascination with superstrings: "The theory depends for its existence upon magical coincidences, miraculous cancellations and relations among seemingly unrelated (and possibly undiscovered) fields of mathematics. Are these properties reasons to accept the reality of superstrings? Do mathematics and aesthetics supplant and transcend mere experiment?"

Glashow's condemnation centered on the old problem that the theory apparently cannot be tested. The scale of superstrings—seventeen orders of magnitude smaller than any phenomena observable with even the biggest particle accelerators—places them beyond the current limits of experimental physics. "Superstring theory," according to Glashow and Ginsparg, "unless it allows an approximation scheme for yielding useful and testable physical information, might be the sort of thing that Wolfgang Pauli would have said is 'not even wrong.'" At the close of a physics conference in 1986, Glashow summed up his feelings in a bit of doggerel verse: "The theory of Everything, if you dare to be bold / Might be something more than a string orbifold. / While some of your leaders have got old and sclerotic, / Not to be trusted alone with things heterotic, / Please heed our advice that you too are not smitten— / The Book is not finished, the last word is not Witten."

Even Witten himself might agree with that assessment. Toward the end of the 1980s, he acknowledged that superstring theory still faced some daunting problems. Hard evidence, such as the discovery of one of the superpartner particles that are required by supersymmetry, would certainly be an encouraging development. However, although some superpartners might lie within the range of planned accelerators, even a successful discovery of such particles would constitute only the flimsiest of links between the real world and the grand theoretical structure of superstrings. Most superstring advocates feel that further progress still depends on a major conceptual breakthrough. Witten and others have thus turned to exploring kindred areas of research in the hopes of coming upon the necessary crucial insight or mathematical clarification.

Whatever the outcome of all this theoretical toil—whether string theory or one of the other approaches or perhaps some entirely unforeseen combination of ideas succeeds—it seems inevitable that physicists will have to accept a radically altered view of what space, time, and matter are. Until that new perspective and the ultimate, all-encompassing Theory of Everything emerge, physicists seem intent on casting their nets wide. As Feynman once put it, "The only thing that's dangerous is that everybody does the same thing. We ought to be running around in as many directions as possible."

Physics is at once the narrowest and broadest of sciences: It seeks, almost as a holy grail, a unified piece of mathematical legislation that can account for all of natural reality, from the emanations of an atom to the evolution of the cosmos. Over the course of the twentieth century, scientists have made measurable strides toward finding such a unified scheme, but until recently, one major piece of the puzzle refused to fit.

Two great postulates, quantum theory and classical general relativity, have provided the underpinnings of modern physics. Quantum theory describes the interplay of the strong and weak nuclear forces and the electromagnetic force in the world of fundamental particles; classical general relativity explains the gravitational force in terms of the geometry of space and time. Each has proved to be a powerful predictive tool in its own right. But attempts to apply the equations of quantum theory to the force of gravity have led to much mathematical grief; among other misfires, the calculations have spawned a blizzard of "infinities"—nonsensical results like those that occur when one is divided by zero.

In the past few years, however, physicists have grown excited about a new breed of entities called superstrings, unimaginably small, one-dimensional structures that may represent the deepest architecture of the universe. Although still hypothetical and incompletely understood, superstrings seem to clear the way for a quantum theory of gravity, leading some enthusiasts to dub the new construct "The Theory of Everything."

Of World Lines and World Sheets

Quantum theory views fundamental particles as mathematical points in space. Flickering into and out of existence and conveying the strong, weak, and electromagnetic forces, they are seen as utterly dimensionless. Physicists long assumed that gravity, too, would be describable as point particles. But when viewed in these point-particle terms, the laws of general relativity transformed space-time from a smooth continuum into a seething foam, possibly accounting for the nonsensical infinities generated by efforts to unify the two great theories.

In the last two decades, however, physicists developed what seemed like a miracle cure: When quantum points are replaced by one-dimensional strings, with length but no thickness, mathematical maladies seem to vanish. Two types of strings have been proposed, an open variety, with free ends, and a closed version—a loop. In both cases, the strings are incredibly small, just 10^{-33} centimeter in length. Their size compares to an atom as an atom compares to the Solar System. Today, most physicists favor schemes with only closed strings because the equations yield more realistic results.

As string theory evolved, physicists incorporated another radical idea that had surfaced at about the same time. Called supersymmetry, it mathematically relates the two classes of fundamental particles—bosons, or force particles, and fermions, or matter particles *(pages 26-27)*. Strings were endowed with the same ability and became known as superstrings.

The difference between point particles and strings is evident in the way they move through space-time, although both move in the manner that expends the least energy—the least-action principle, as it is known. A point particle follows the shortest distance between two locations, called a "world line." But because a string has length, it sweeps out a "world sheet." World sheets always have the smallest area possible in space-time. The laws of physics allow a string to get longer or shorter over time, and to vibrate much the same way a violin string does. Each vibrational frequency gives rise to a large collection of fundamental particles *(pages 82-83)*.

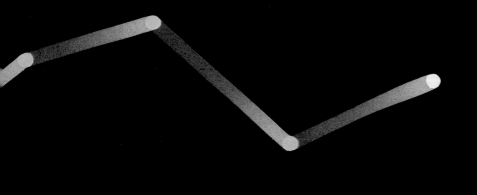

Point particle. The world line of a point particle traces out a one-dimensional path through both space (represented on the vertical axis here) and time (horizontal axis). The path, following the least-action principle, is the shortest distance between two points in space-time.

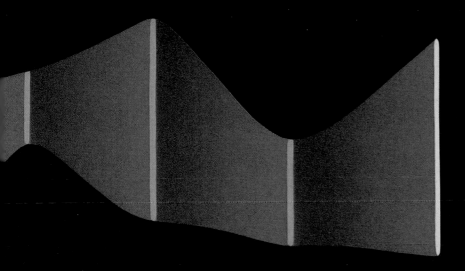

Open string. An open string sweeps out a world sheet as it moves through space-time. Like a point particle, an open string moves with maximum economy and sweeps out the smallest area possible. Distortions in the sheet along the string's vertical axis have no physical importance, but vibrations perpendicular to this surface appear physically.

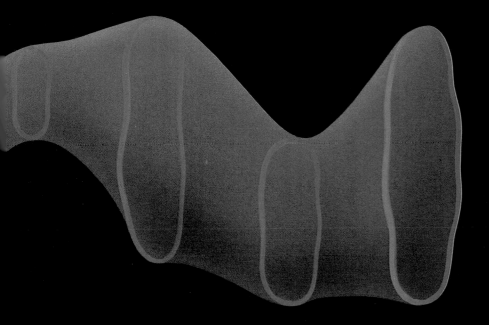

Closed string. A closed, or looped, string has no endpoints and sweeps out a tubelike world sheet. The tube is similar to a bubble created when a child sweeps a loop filled with soapy water through the air.

A Vibrational Medley

According to superstring theory, all the fundamental particles that make up the universe are manifestations of string vibrations and the properties inherent in them. As with a violin string, the vibrations of these infinitesimally small, one-dimensional curves follow strict mathematical rules, akin to those of the har-monics of music. Different degrees of excitation, known as vibrational states, engender distinct sets of subatomic particles whose masses are precisely determined. The higher the vibrational state *(box, bottom)*, the greater the mass of the particles in the set created by the string's oscillation.

As illustrated below and on the opposite page, all of the fundamental particles observed in our universe—quarks, electrons, photons, and so on—are assumed to

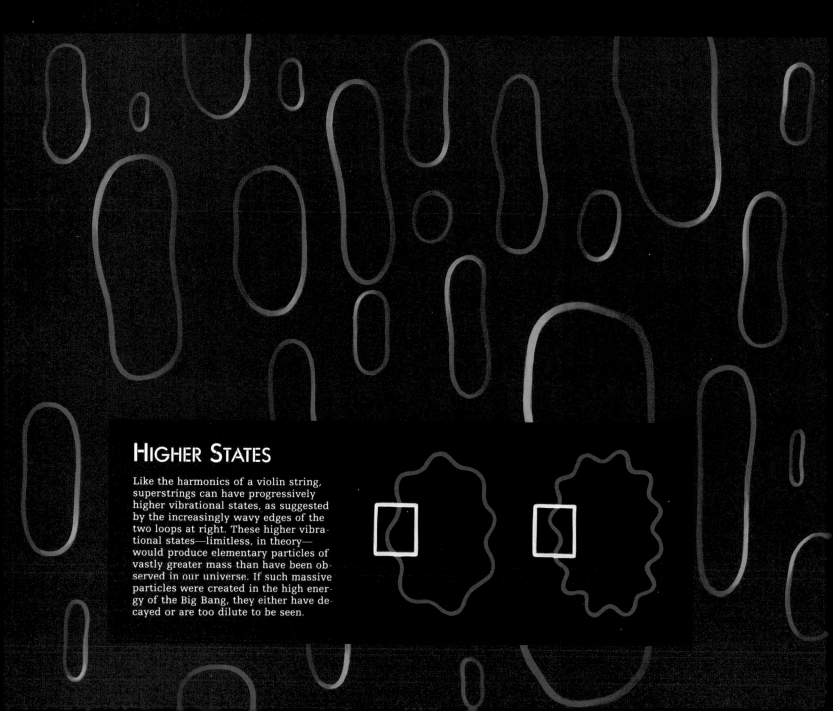

Higher States

Like the harmonics of a violin string, superstrings can have progressively higher vibrational states, as suggested by the increasingly wavy edges of the two loops at right. These higher vibrational states—limitless, in theory—would produce elementary particles of vastly greater mass than have been observed in our universe. If such massive particles were created in the high energy of the Big Bang, they either have decayed or are too dilute to be seen.

come from the lowest vibrational state, called the zero mode. This presents a substantial difficulty for physicists, since superstring theory implies that the minimal mode should produce particles with no mass at all, which is clearly not the case. Given the embryonic condition of the theory, however, scientists are hopeful that the solution lies in wait.

Superstring theory states that the mass of the sets of particles increases without limit and each increase will be of great magnitude. For example, the vibrational state immediately above the zero mode would produce a set of particles with masses approximating that of a mote of dust, behemoths compared to any known elementary particle. The next note up the superstring scale would yield a set with masses twice that. Whether such higher orders of mass ever existed is unknown—one of many mysteries still attached to this radical new concept of a string-based cosmos.

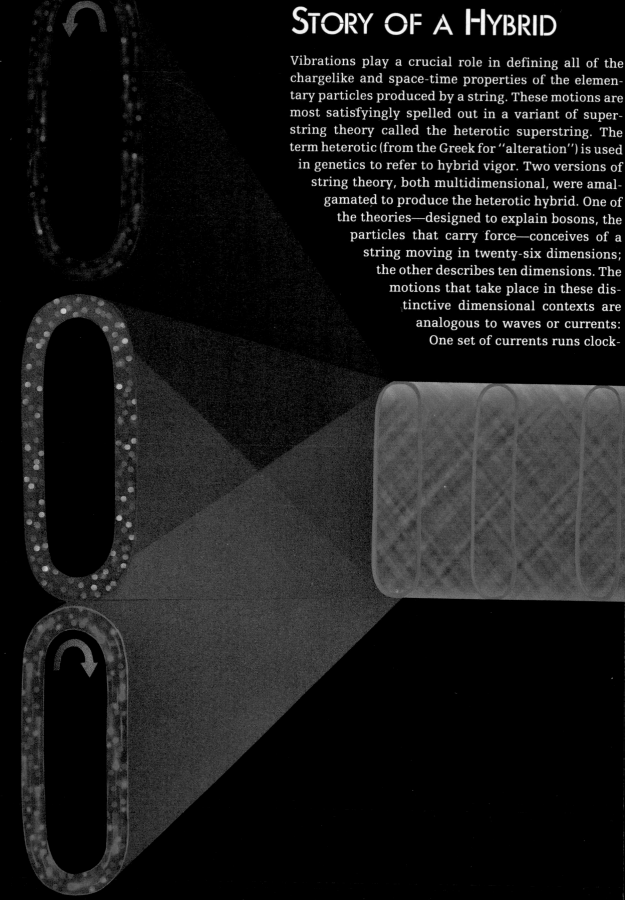

STORY OF A HYBRID

Currents running counterclock-wise on a four-dimensional heterotic superstring account for all the observed spin and generational properties of particles. Spin provides the distinction between bosons and fermions; generational properties define specific fermions—members of the lepton and quark families.

Vibrations play a crucial role in defining all of the chargelike and space-time properties of the elementary particles produced by a string. These motions are most satisfyingly spelled out in a variant of superstring theory called the heterotic superstring. The term heterotic (from the Greek for "alteration") is used in genetics to refer to hybrid vigor. Two versions of string theory, both multidimensional, were amalgamated to produce the heterotic hybrid. One of the theories—designed to explain bosons, the particles that carry force—conceives of a string moving in twenty-six dimensions; the other describes ten dimensions. The motions that take place in these distinctive dimensional contexts are analogous to waves or currents: One set of currents runs clock-

Standing currents (combinations of clockwise and counterclockwise currents) generate the four-dimensional properties of familiar space-time. The standing waves also account for some of the properties of the graviton, the theoretical particle that carries the gravitational force.

The clockwise currents of a four-dimensional heterotic superstring carry the charge properties corresponding to the fields of the weak, strong, and electromagnetic forces. These currents thus dictate the characteristics of the force carriers—intermediate vector bosons, gluons, and photons, respectively.

wise around the heterotic string; another runs counterclockwise. To reconcile the heterotic superstring with the four dimensions of classical relativity, physicists propose standing waves or currents that combine attributes of both currents. As shown here, the four-dimensional heterotic string generates different particles by the joint action of these currents, causing a heterotic string to manifest itself as an up quark, a muon, a neutrino, or any of scores of other creatures in the subatomic bestiary.

Although a theory that not only requires many more dimensions than our familiar four-dimensional spacetime but also calls for two mismatched systems of dimensions may seem inordinately complicated, superstring theorists point out that the mathematics of the heterotic model work extraordinarily well—to the extent that they have been fathomed. In any event, the multiple internal dimensions may be something else, not dimensions in the usual sense but simply a way of describing internal charge conditions.

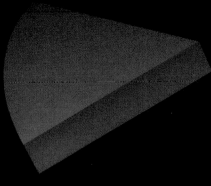

Certain properties of quarks—fundamental constituents of matter that come in six known "flavors," have a spin of ½, and possess a property called "color" that is similar to electric charge—arise from a combination of all three currents of a heterotic superstring.

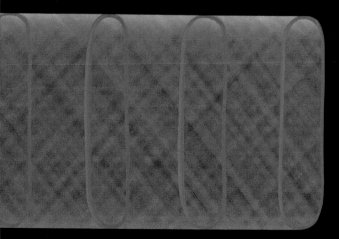

The properties of the force-carrying bosons—in this case, a photon, with no mass, no charge, and a spin of 1—are determined by a different combination of the three currents.

Gravitons—the hypothetical gravity-bearing particles, with a spin of 2, zero mass, and no charge—are largely produced by a superstring's standing currents.

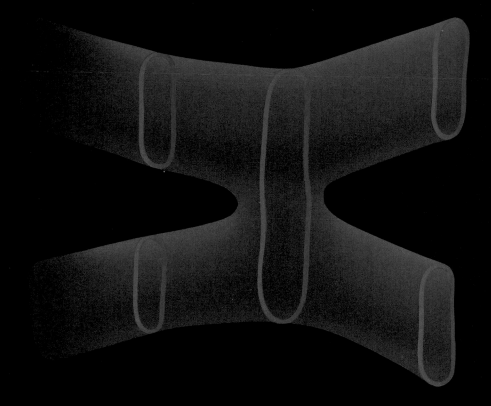

The pants diagram is a tubelike world sheet that portrays two heterotic strings joining into one string, then splitting apart as they move through time.

AN INTERPLAY OF LOOPS

Whereas quantum theory imagines particle interactions as a complex set of collisions and exchanges, superstring theory represents them as a straightforward joining and splitting of loops. The merger and subsequent breakup of two strings is seen above in a kind of illustration that physicists have dubbed a "pants" diagram for its resemblance to a pair of trousers. This sequence is fundamental to heterotic string theory and can account for all particle interactions, including those engendering the basic forces of the universe. In the illustrations at top right, those four forces are portrayed in the traditional terms of point particles, a view that involves an exchange of bosons. A joining and splitting of superstrings incorporates the same information and more.

The pants diagram of string interactions allows physicists to simplify the representation of events that are extremely complex. When two strings join and separate, then join again, a hole is left at the center of the diagrammatic plot of their history. With still more interactions, more holes are created, which leads to a staggeringly intricate structure that seems to realistically represent the quantum-mechanical behavior of the universe.

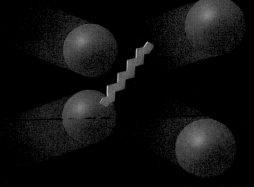

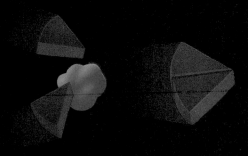

The electromagnetic force—responsible for atomic structure, chemical reactions, and electromagnetic phenomena—is depicted above in point-particle terms, with two electrons exchanging a photon, then scattering.

The strong force, which holds the nuclei of atoms together, has great strength over very short distances. It is represented here by two quarks exchanging a gluon—the carrier of the strong force—and drawing together.

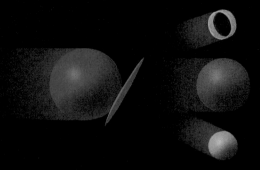

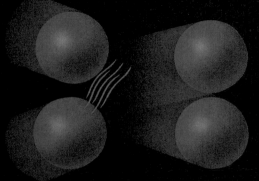

In a conventional quantum theoretical view of the weak force in operation, a muon emits an intermediate vector boson that instantly decays into an electron and an antineutrino, as the muon transforms into a neutrino.

In a gravitational interaction as envisioned by quantum theory, two muons approach each other and exchange a graviton, with a tiny effect that draws them slightly closer together.

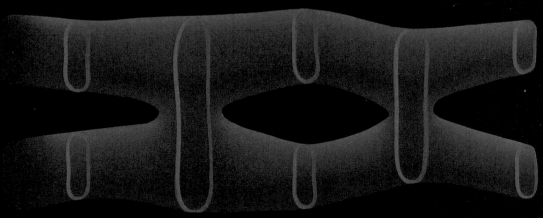

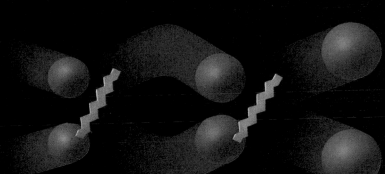

Above, two strings moving to the right through time join and split, then join and split again, producing a hole in the world sheet created by their interactions. At right is a point-particle equivalent of the diagram—two electrons exchange a photon, scatter, then meet to exchange another and scatter again.

Albert Einstein's theory of general relativity shows space as a two-dimensional sheet *(below)* that is warped by the mass of the objects that rest upon it—the degree of warping depending on the amount of mass. The laws of classical physics require that anything moving freely across this sheet must follow the shortest possible route. Because of the warping, the path would be curved.

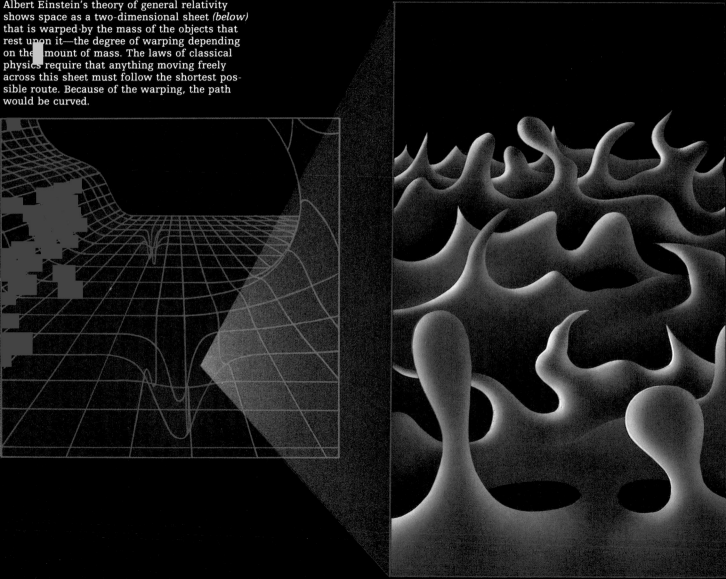

derstanding. Strings exist at what is known as the Planck scale, 1/100,000,000,000,000,000,000 the size of a proton. Joining and splitting in uncountable profusion, they generate prodigiously complicated world sheets, like that suggested in the large illustration below. The interplay of the strings causes fingers to materialize and disappear, an activity that describes, among other things, the ceaseless appearance and disappearance of gravitons, the quantum particles of gravity. Unlike the space-time foam of general relativity, this string vision of gravity in the quantum world makes mathematical sense; it arises from the same rules that underlie everything else in nature.

But the rules of superstring theory are far from being thoroughly understood. For example, the theory calls for strings to create their own space-time rather than splitting and joining in some background space-time. Physicists have not fully worked out how this would occur. Indeed, superstrings pose a mathematical challenge that may not be overcome for decades. In a sense, science is not yet equipped to probe the deeper layers of this remarkably compelling theory.

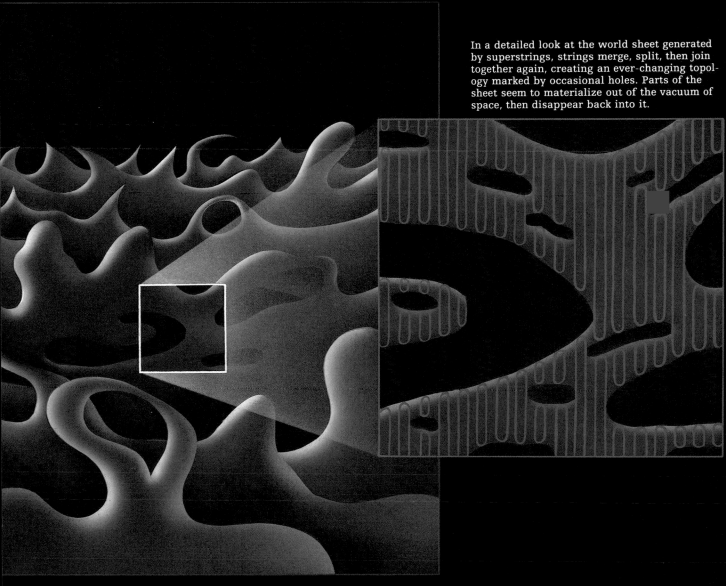

In a detailed look at the world sheet generated by superstrings, strings merge, split, then join together again, creating an ever-changing topology marked by occasional holes. Parts of the sheet seem to materialize out of the vacuum of space, then disappear back into it.

The inner works of a sixteenth-century clock were designed to measure that elusive phenomenon: the passage of time. In twentieth-century physics, time came to be understood as an elastic dimension, inseparable from space in the larger workings of the universe.

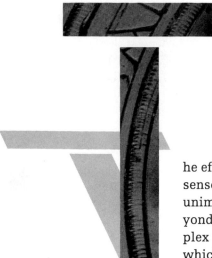

he effort to unify the fundamental forces of nature is, in a sense, an attempt to recapture the past, a past that is unimaginably close to the beginning of the universe, beyond which time itself ceases to have meaning. Using complex formulas that include such factors as the rate at which the universe is thought to be expanding and the intensity of background radiation attributable to the Big Bang, scientists have traced the history of the cosmos back to within 10^{-43} second of its inception—less than a millionth of a trillionth of a trillionth of a trillionth of a second. At that point, calculations suggest, the universe was so dense and hot that energy and matter were indistinguishable and three of the four fundamental forces—electromagnetism, the strong force, and the weak force—might have been fused. But no theorist has yet been able to unite the relatively feeble action of gravity with the other three forces even at the intense energies projected at that instant. And any effort to look back before that moment runs up against a forbidding barrier: the incalculable effects of a universe approaching infinite density. According to cosmologist Steven Weinberg, time in a superdense universe may be akin to a temperature of absolute zero. "We may have to get used to the idea of an absolute zero of time," he notes, "a moment in the past beyond which it is in principle impossible to trace any chain of cause and effect. The question is open, and may always remain open."

That scientists can even contemplate such a conundrum is a tribute to centuries of painstaking effort that helped quantify the once-mysterious relationship between time and space. Theories such as Einstein's general relativity, which are based in part on the principle that light has a constant, and finite, speed, have provided the framework for theorists seeking to describe the earliest moments of the universe. Yet determining the speed of light (the rate at which all electromagnetic energy travels through space over time) was a labyrinthine problem that challenged researchers for more than two centuries, from Newton's day through Einstein's. And in any case, even the most recent and precise such calculations have done little to dispel the fundamental paradoxes of time. Theorists today still ponder the age-old question of whether time on the cosmic scale is like a line or a circle—an irreversible progression from birth to extinction or an endless round of expansions and contractions. To consider such problems is to take up matters that have perplexed curious minds since antiquity.

"WHAT, THEN, IS TIME?"

Saint Augustine of Hippo, the preeminent philosopher of the early Christian Church, never shrank from an intellectual challenge, and one of the most difficult he ever accepted was to explain the meaning of time. Around the year 400, in the late days of the Roman Empire, Saint Augustine bent to this self-appointed task in his autobiographical meditation, the *Confessions.* He began by considering the provocative question: "What was God doing before He made heaven and earth?" Augustine dismissed the sly retort sometimes offered to this riddle—that "God was preparing Hell for people who pry into mysteries." The issue deserved very serious consideration, he believed, for it raised the disturbing possibility that a seemingly purposeful God either had idled away an eternity before the Creation or, perhaps even more disturbing, had created and disposed of other things before getting around to Earth and humankind.

To get at this problem, Augustine first tried to define the elusive entity called time. He noted that between the future and the past, neither of which can be said to exist except in the imagination, lies a fleeting present that slips away before the intellect can grasp it. "What, then, is time?" he asked. "I know well enough what it is, provided that nobody asks me; but if I am asked what it is and try to explain, I am baffled."

Characteristically, Saint Augustine did not rest there, but went on to examine various ways in which other philosophers had described time and studied it. He looked closely, for example, at the proposition that "time is nothing but the movement of the sun and the moon and the stars." After all, the orbits of the Sun and Moon and the wheeling of the constellations about the heavens had long been used to mark the passage of time, so it seemed reasonable to identify time with such cycles of nature. Augustine pointed out, however, that the movements of heavenly bodies are not necessarily consistent with the orderly conception of time as it is expressed by such fixed intervals as hours, days, and months. As evidence for this proposition, he invoked a biblical precedent—the tale of Joshua calling on God to stop the Sun in its tracks so that the Israelites would have extra daylight to defeat their enemy; according to Scripture, the Sun "stood still in the middle of the sky and delayed its setting for almost a whole day." But Augustine might just as well have cited the problems that the Romans and other ancient peoples encountered in attempting to draw up calendars that reconciled lunar, solar, and sidereal cycles.

In distinguishing the flow of time from the movement of heavenly bodies, Augustine was echoing an earlier authority, the Greek philosopher Aristotle, who had observed that while time can be gauged by the motion of the Sun and Moon, those celestial motions and others on Earth can in turn be measured by the passage of time. Aristotle concluded that in some way motion and time must be separate phenomena, since they can be used independently to monitor events. Yet he also believed that time was rooted in nature, and he clung to the traditional view that the best symbol for time is a circle, whose shape evokes the prospect of the future leading back to the past just as the sunset leads to the dawn.

This cyclical view of time, common among ancient cultures, was anathema to the early Christians, who regarded human experience as a one-way journey from Genesis to the Judgment Day. Saint Augustine reinforced that point, and departed from Aristotle, by rejecting any connection between time and nature's recurring patterns. Indeed, he came to the conclusion in the *Confessions* that time was conceived by God at the Creation—there was no time to fill before that—and that the human mind could follow time's inexorable progress without reference to natural cycles, through the God-given gifts of memory and expectation. In a later work, *City of God,* Saint Augustine argued

Shadow tracking devices, such as this Egyptian sundial from about the third century BC, were the first attempts to measure time's passage. The shadow of a rod placed in the round hole moved over lines marking twelve hours as the sun crossed the sky.

At night and on cloudy days, Egyptians relied on water clocks similar to this model. A hole near the bottom let the water level fall past twelve marks on the inside of the bowl. A seventy-degree slant to the bowl's sides kept the clock accurate to within ten minutes a day.

SUBDIVIDING TIME

Nature gives the world daylight and darkness, in measures that vary with the season. Early on, however, the increasing complexity of human affairs began to require more precision and smaller units of measurement. Hence, the invention of the hour, a twelfth of the day's sunlight. On these and the following pages are some of the major milestones in timekeeping, from the sundial *(above)* to the atomic clock *(page 97)*—a procession that reflects centuries of changing ideas about the nature, and divisibility, of time itself.

that the notion of circular time had been put forward by "deceived and deceiving sages." Past events would not be repeated in the future, he emphasized. Instead, the world would move triumphantly forward, progressing from the corruption of Rome toward the day when true believers would be gathered into the kingdom of God.

Saint Augustine thus helped propound the concept of time as a steady progression of events running from the past to the future. Although inspired by faith, this abstract approach would contribute to the scientific thinking of a later age. But significant progress on the experimental level would depend on practical breakthroughs—the invention of better instruments for measuring time as it proceeded along its presumably inexorable course.

Medieval sandglasses used the flow of sand through a narrow neck to mark the start and end of prayers, lessons, or a watch aboard ship. In the eighteenth-century German glasses shown below, sands of various grades of fineness run out in *(left to right)* fifteen, thirty, forty-five, and sixty minutes.

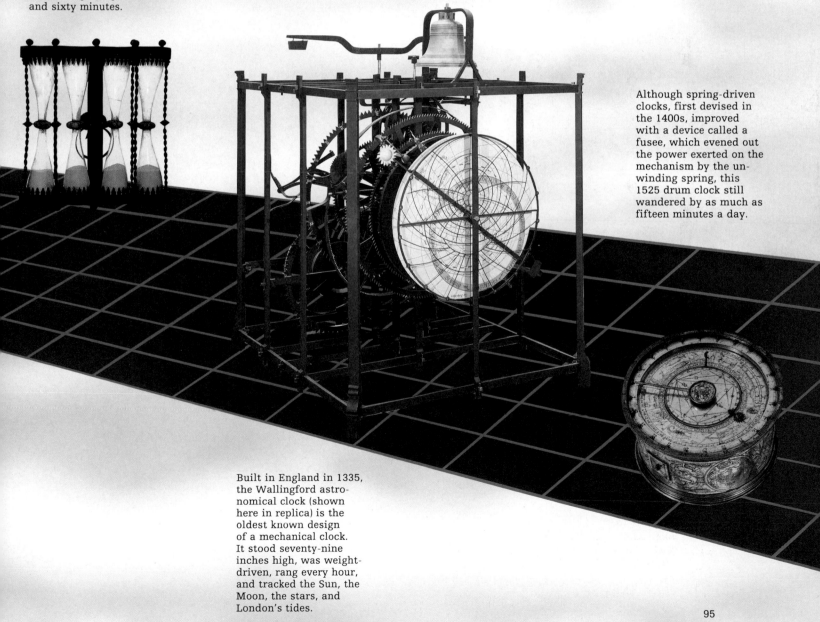

Although spring-driven clocks, first devised in the 1400s, improved with a device called a fusee, which evened out the power exerted on the mechanism by the unwinding spring, this 1525 drum clock still wandered by as much as fifteen minutes a day.

Built in England in 1335, the Wallingford astronomical clock (shown here in replica) is the oldest known design of a mechanical clock. It stood seventy-nine inches high, was weight-driven, rang every hour, and tracked the Sun, the Moon, the stars, and London's tides.

95

THE ART OF TELLING TIME

Timekeeping was far from precise in ancient eras. To be sure, many civilizations used sundials or other kinds of shadow clocks to mark the passage of the day. With the help of such devices, the Sumerians divided the period of daylight into twelve segments; the Egyptians later did the same with the night by gauging the movements of stars, thus instituting the twenty-four-hour day. But the duration of those hours varied; the daylight hours, for example, were shorter in the winter than in the summer, when the twelve intervals inscribed on Egyptian shadow clocks divided up a longer period of sunlight. By the rise of the Roman Empire, the length of an hour was set at one twenty-fourth of the duration of a full day, allowing for the use of slow-dripping water clocks and sand-filled hourglasses to mark off such intervals. But those instruments had to be turned or refilled promptly and frequently, and even with the best of care the time indicated by one could differ substantially from that denoted by another.

Improvements came slowly. By the eleventh century AD, the Chinese had perfected a long-running clock that incorporated a waterwheel, with cups affixed to the rim. A spout fed a consistent flow of water into each cup in turn; when one was full, it descended by virtue of its own weight and moved the

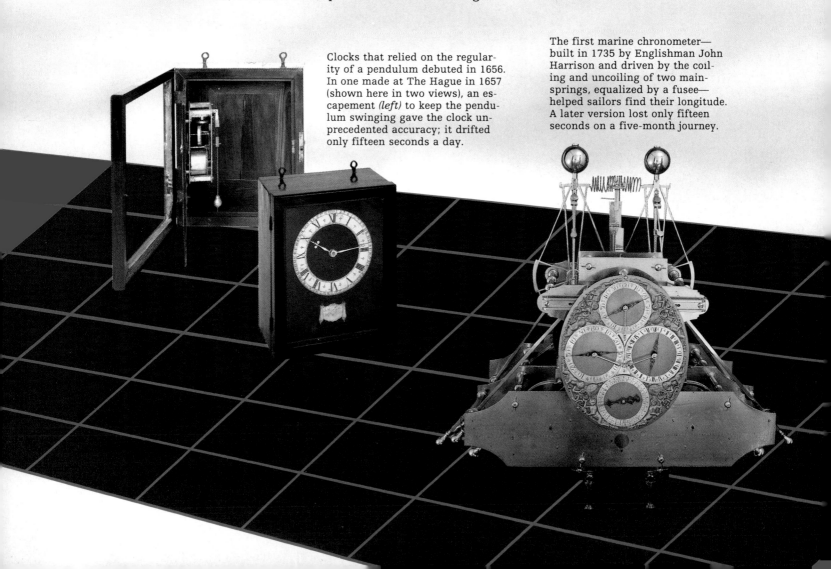

Clocks that relied on the regularity of a pendulum debuted in 1656. In one made at The Hague in 1657 (shown here in two views), an escapement (left) to keep the pendulum swinging gave the clock unprecedented accuracy; it drifted only fifteen seconds a day.

The first marine chronometer—built in 1735 by Englishman John Harrison and driven by the coiling and uncoiling of two mainsprings, equalized by a fusee—helped sailors find their longitude. A later version lost only fifteen seconds on a five-month journey.

wheel along another notch, ticking off time. However ingenious, such mechanisms were hard to build and to maintain, and were as useless on a freezing day as sundials were on a cloudy one.

The first reliable mechanical clocks appeared in Europe in the thirteenth century, and were housed in monastery towers. They had no hands, but their revolving wheels told monks when to toll bells for prayers. Like earlier clocks, they exploited the action of gravity, but in place of flowing water or sand, they relied on a falling weight that was connected to a wheel by a rope wound about the wheel's axle like a string around a top; the rim of the wheel had teeth, which were engaged by a balanced device known as an escapement that broke the weight's descent into small steps, so that the wheel, or gear, revolved one tick at a time. Such clocks all required periodic expenditures of human energy—someone had to crank the weight back up—and they erred by as much as an hour a day. But they functioned in all seasons, and the gears provided

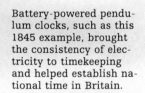

Battery-powered pendulum clocks, such as this 1845 example, brought the consistency of electricity to timekeeping and helped establish national time in Britain.

The first atomic clock, built in 1948 by the U.S. National Bureau of Standards, harnessed the resonant frequency of ammonia molecules to a quartz crystal, keeping the clock accurate to one second in three years.

During the late 1960s, the accuracy of quartz crystals came to the private citizen in battery-powered wrist watches such as this 1968 model, made in Japan by Seiko. Today, such timepieces gain or lose about one minute per year.

a convenient way of dividing time into short intervals. In the fourteenth century, minutes and seconds were defined, and hour and minute hands became commonplace on clocks.

A fascination with timekeeping spread across Europe. Every town with the funds, it seemed, wanted to erect a large public clock as a point of civic pride. In 1382, Philip the Bold of Burgundy capped a victory over his Flemish opponents by entering the town of Courtrai, hauling down its belfry clock, and carting it back to his capital at Dijon. The vogue for such showpieces literally put time on a pedestal and drew wide attention to the movement of the hour and minute hands. Since the days of hourglasses, it had been possible to see time slipping away, but now the process was inextricably linked to a numerical sequence, instilling in the mind a sense of the mathematics of time that helped pave the way for modern science.

CONNECTING TIME AND SPACE

Although various scholars of the late Middle Ages speculated on the relationship between time and space as expressed by the motion of objects, it took a practical genius to begin quantifying that relationship. The ceaselessly inventive Italian scientist Galileo Galilei made that leap in the early seventeenth century with the help of a device whose value for timekeeping he was apparently the first to recognize—the pendulum. According to his student and biographer, Vincenzo Viviani, Galileo realized the potential of the pendulum in a serendipitous way, by observing lamps swinging from the ceiling of the cathedral in Pisa during an earthquake. Timing the to-and-fro movements with his pulse, he found their period to be remarkably regular. After his death, the pendulum would be used to regulate a new type of escapement for clocks that would greatly improve their accuracy. But during his lifetime, Galileo pioneered the use of pendulums and other timekeeping devices as experimental tools.

One of his greatest experimental feats was to measure the rate at which objects fall. Other curious observers had noticed that objects dropped from a height appear to speed up as they approach the ground, but Galileo set for himself the goal of determining "the proportion according to which this acceleration takes place." The task demanded more than an accurate timepiece. Since speed equals distance traveled over time elapsed, Galileo had to devise a way of determining how far an object fell in a given interval of time. To that end, he made ingenious use of a water clock to measure the time it took a ball to roll down various lengths and slopes of an inclined plane *(opposite)*. Finding that the ratio of the weight of water that escaped with each trial was, in effect, a ratio of the time, he concluded that the speed of a falling object is proportional to the time elapsed. That is, if an object is traveling at thirty-two feet per second at the end of the first second of its fall, it will move twice as fast—sixty-four feet per second—at the end of the next second, and three times as fast at the end of the third.

This discovery laid the groundwork for Isaac Newton's brilliant elaboration of his universal law of gravitation, which was spelled out in 1687 *(page 13).* To be sure, Newton, like Galileo before him, was aware of the difficulties that sometimes arose in trying to define the distance that an object travels over time. Newton cited the case of a sailor walking along the deck of a moving ship. The sailor would be moving at one speed relative to the ship, and at another speed relative to a fixed object such as an anchored buoy. But Newton was convinced that once a spatial frame of reference is settled upon, the problem of relativity disappears, because time is invariable. "All motions may be accelerated or retarded," he insisted, "but the flowing of absolute time is not liable to any change."

Newton's faith in the regularity of time reflected both the mechanistic outlook of his age, in which clocks were becoming increasingly dependable, and the philosophical view that Saint Augustine and others had bequeathed to Western culture—the idea of time as a steady course that never meanders or doubles back on itself. Indeed, one technique Newton used to analyze motion was to represent time as a straight line and distance as a second axis running perpendicular to it. An object moving at a constant speed would be represented on this grid as a straight line, while one whose speed was changing would be represented as a curve. But Newton did not confine himself to

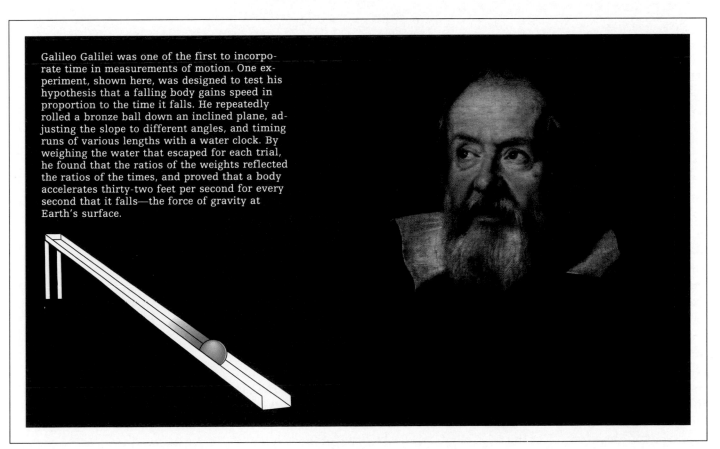

Galileo Galilei was one of the first to incorporate time in measurements of motion. One experiment, shown here, was designed to test his hypothesis that a falling body gains speed in proportion to the time it falls. He repeatedly rolled a bronze ball down an inclined plane, adjusting the slope to different angles, and timing runs of various lengths with a water clock. By weighing the water that escaped for each trial, he found that the ratios of the weights reflected the ratios of the times, and proved that a body accelerates thirty-two feet per second for every second that it falls—the force of gravity at Earth's surface.

In a celebrated eighteenth-century feud, English mathematician Isaac Newton *(left)* and German philosopher Gottfried Wilhelm Leibniz disputed, among other things, the nature of time. Newton held that time was an absolute, "flowing equably without relation to anything external." Leibniz argued that time was merely an abstraction, invented to help humans order a succession of events. Newton's idea prevailed until the advent of Einstein's theories of relativity.

charts. He developed a new branch of mathematics to analyze situations where a function such as speed changes over time. To determine speed at any instant amid such flux, Newton in effect sliced the time intervals into infinitesimal segments. He called this new technique "fluxions," but it soon became known as calculus.

Calculus was not an invention that Newton could stake sole claim to. In a striking coincidence, another great mathematician of the day, German Gottfried Wilhelm Leibniz, independently devised the same system. Leibniz published his account of calculus first—in 1684. Nonetheless, Newton insisted that he had worked out the technique earlier, though he did not take the trouble to publish his work until more than a decade later. A bitter dispute arose, not only over which of the two scientists was the true father of calculus, but over the very nature of time and space. Newton and Leibniz never met to contest the issue. Instead, the debate was carried on through intermediaries. In 1715, Leibniz presented his side of the story in a letter to Caroline, Princess of Wales, whom he had first met when she was a youngster staying at the court of Frederick I in Berlin. Caroline showed the letter to one of Newton's champions, Samuel Clarke, a theologian at Westminster. Clarke gave reply in a long letter defending Newton. Thus began a two-year correspondence between Leibniz and Clarke that was later published.

In his letters, Leibniz attacked Newton not only on scientific but on religious grounds, complaining that Newton's concept of time reduced God to the role of a watchmaker whose interventions amounted to winding up the clock on occasion or mending it. Leibniz believed that Newton had turned time and space into idols, treating them as absolute and eternal when they were in fact nothing but figments—imaginary strings that held together real events and objects. Clarke responded that time and space were not mere arrangements of things but quantities that could be used to measure the intervals between things. If time was nothing but a succession of events, he pointed out, then there was no timely way to relate that succession to what preceded it: "It would follow that if God had created the world millions of ages sooner than He did, yet it would not have been created at all the sooner." Thus the argument circled back to the issue that had perplexed Saint Augustine: the meaning of time at the beginning of things.

This lofty debate had little practical impact, for most scientists were less impressed by the philosophical differences between Newton and Leibniz than by the splendid tool they developed independently for studying changes over time. In the long run, however, neither calculus nor improved experimental techniques could dispel the deeper riddles of time and space. To the contrary, careful empirical studies eventually brought to the fore apparent exceptions to the supposedly universal laws of motion that could be addressed only by confronting such basic issues as whether time was indeed absolute. The most perplexing of these exceptions involved the motion of light, whose speed was already being gauged in rough fashion while Newton and Leibniz were devising a new language for physics.

MEASURING THE SPEED OF LIGHT

Before the scientific revolution, natural philosophers had assumed that the speed of light was infinite. The ever skeptical Galileo tried to test that assumption in the early 1600s equipped only with an assistant and a pair of lanterns. He asked his assistant to stand on a distant hill and wait for a signal from a lantern, then return the flash at once so that Galileo could measure the time it took the light to cover the distance to the hill and back. All the experiment revealed was that light travels extremely fast, much faster than the reflexes of people waving lanterns.

Better results were obtained in the 1670s by Ole Römer, a Danish astronomer who studied the eclipses of the four moons Galileo had discovered

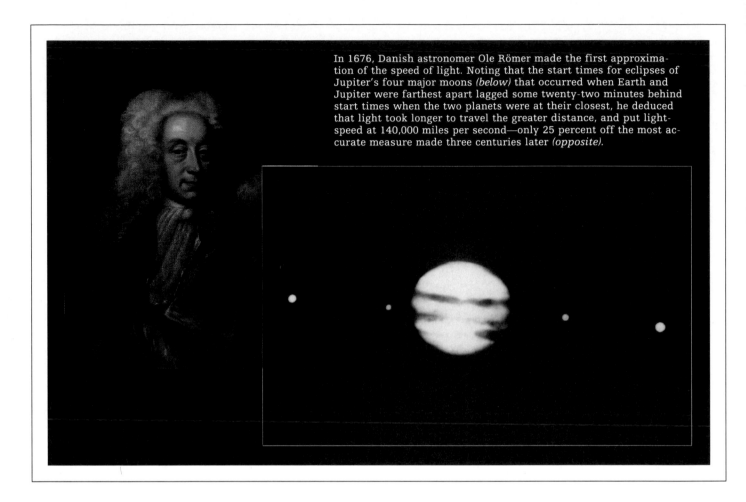

In 1676, Danish astronomer Ole Römer made the first approximation of the speed of light. Noting that the start times for eclipses of Jupiter's four major moons *(below)* that occurred when Earth and Jupiter were farthest apart lagged some twenty-two minutes behind start times when the two planets were at their closest, he deduced that light took longer to travel the greater distance, and put light-speed at 140,000 miles per second—only 25 percent off the most accurate measure made three centuries later *(opposite)*.

around the planet Jupiter. Timing the periodic disappearances of the moons as they circled behind the great planet, Römer noticed something strange: The eclipses began progressively later as Earth moved away from Jupiter. The difference amounted to about twenty-two minutes between the night on which Earth and Jupiter were nearest each other and the night, six months later, when Earth had traveled through half its orbit and the two planets were farthest apart. Römer astutely attributed this lag to the extra distance that light from the moons had to travel. Because Jupiter completes just one twenty-fourth of its orbit during the six-month period, a distance that could be largely discounted, Römer concluded that the extra distance amounted to the diameter of Earth's orbit.

Thanks in part to Newton's studies of orbital dynamics, Römer had a rough idea of how wide the diameter of Earth's orbit was, and the ratio of that distance to the time it took light to span it—twenty-two minutes—yielded an approximate speed for light of 140,000 miles per second. It was not a bad approximation, falling about 25 percent short of the actual figure. Although the announcement did not greatly impress Römer's contemporaries, many of

In 1983, several teams of physicists, including *(left to right)* Don Jennings, Russell Petersen, and Ken Evenson at the National Institute of Standards and Technology in Boulder, Colorado, used sophisticated laser equipment to help determine the speed of light with unsurpassed accuracy. By measuring the frequency of a helium-neon laser beam—473 trillion cycles per second—and multiplying that value by the wavelength of the same laser, the teams calculated that light travels 186,282 miles per second.

whom clung to the belief that the speed of light was beyond measure, later scientists would regard Römer's feat as a turning point. In the words of the twentieth-century British astronomer Sir Arthur Eddington, the measurement marked the beginning of "time as we know it."

FINE-TUNING

Some fifty years after Römer offered his figure for the speed of light, Englishman James Bradley, an astronomer at Oxford University, improved on it. Bradley was interested in measuring stellar parallax, or the apparent shift in the location of a nearby star against the background of very distant stars, as observed from a point on Earth whose position relative to the star changes with the planet's orbital movement around the Sun. In 1725, Bradley attempted to trace the parallax shift of the star Gamma Draconis. He did detect an apparent shift in the star's position, but it could not be explained in terms of parallax (as astronomers later determined, Gamma Draconis is so far away from Earth that the movement of the planet from one side of the Sun to the other could have no discernible effect on the star's observed position).

Bradley puzzled over the matter for some time. One day, as he was sailing on the Thames, he noticed how the wind vane on the mast of the boat shifted direction as the vessel turned about in a constant wind. It dawned on him that something similar was taking place when he observed Gamma Draconis from the moving Earth. Light from the star could be compared to a wind moving at a fixed speed and direction, and the orbiting Earth to a turning ship. The telescope, like the vane, would try to align itself with the flow of light, but to do so it would have to account for Earth's orbital motion, just as the direction indicated by the vane reflected the boat's movement.

If, for example, the wind was blowing due south at ten knots while the boat was moving eastward at ten knots, the vane would indicate a wind coming from the northeast. Significantly, the apparent direction of the wind would depend not only on the direction of the two components but on their relative velocity; that is, if the speed of the north wind was instead thirty knots and the boat moved east at ten knots, the wind as perceived on deck would be subject to less deflection and would appear more northerly. In the same way, the stellar aberration Bradley had observed involved the slight deflection of starlight arriving from one direction at a very high velocity when viewed from a point on a planet orbiting in another direction at a lower velocity. Bradley was able to quantify this effect, using a figure for Earth's orbital speed derived from the estimated width of its orbit. He could then fill in the missing value—the speed of light, which he calculated at 183,000 miles per second.

Bradley's ingenious technique was sound, and his estimate—only two percent off today's figure—was generally accepted. But like Römer's calculation, this one depended on astronomical measurements that had a small but significant margin of error. To get around that problem, later researchers attempted to succeed where Galileo had failed and compute the speed of light on the ground. In 1849, Armand Fizeau, a wealthy French scientist who financed his own work, mimicked Galileo's experiment, but with a far more sophisticated lantern—a device that shone light through gaps in the rim of a spinning wheel to a mirror 5.16 miles away that in turn reflected the beam back to the observer. Fizeau increased the spin rate of the wheel until the light no longer returned between the gaps in the wheel but was instead blocked by the intervening teeth, which indicated that light was covering the distance to the mirror and back in the tiny fraction of a second it took the wheel to revolve from gap to tooth.

Fizeau's countryman, Jean Foucault, improved on this technique by replacing the wheel with a rotating mirror of many facets and measuring the time it took for light reflected by one facet to be received back by its neighbor. Experiments with this device yielded the best figure for the speed of light to date—about 186,600 miles per second. Foucault's technique was then adopted and refined by American physicist Albert Michelson, who capped a remarkable career devoted to the study of light in 1931, when he refined his measurements to yield a speed of 186,271 miles per second. This stood as the global standard until the early 1980s, when several groups of researchers, including a team at the U.S. National Bureau of Standards, arrived at a figure of 299,792.4586 kilometers, or 186,282.3974 miles, per second *(page 103)*.

A CRACK IN THE FOUNDATION

Deriving a value for the speed of light that stood for five decades was only a small part of the legacy of Albert Michelson. Early in his career, he pointed theorists toward a problem that would undermine the Newtonian conceptions of time and space and would propel physics in a new and baffling direction. As befit a researcher whose work helped to alter the scientific map, Michelson's background was international. Born in what is now Strzelno, Poland, in 1852, he emigrated with his parents to the United States at the age of two. In 1880, after graduating from the U.S. Naval Academy and teaching science there for a time, he traveled to Europe to study the physics of light in Berlin, Heidelberg, and Paris. Returning to America in 1882 to take up an academic career—he eventually became head of the physics department at the University of Chicago—Michelson embarked on a series of inquiries into the behavior of light that in 1907 earned him the Nobel prize, the first ever awarded to a citizen of the United States.

As it happened, the experiment that secured Michelson's place in history was one that he regarded as a disappointment. In 1887, he and chemist Edward Morley collaborated on a test designed to detect the presence of the so-called ether, the insubstantial but universal medium through which all electromagnetic waves, including visible light, were presumed to propagate. Physicist James Clerk Maxwell had recently established that the transmission of electromagnetic energy could be analyzed in terms of wave action, and it seemed only logical that light waves, like sound waves, must require a medium for transmission.

According to theory, the ether was static and distinct from the gaseous atmosphere that clung to Earth. Just as running through still air causes the runner to feel the breeziness of parting air, observers on Earth should be able to detect the tenuous ether wind moving past as Earth orbits the Sun. And just as sound waves may change speed as they interact with air and water, the same should be true of light as it moves through the ether. Thus, Michelson and Morley set out to detect the ether by measuring its effect on the speed of light propagated at Earth's surface.

The experiment made use of a device Michelson had invented several years earlier. Set on a rotating frame, this interferometer, as it was called, split a beam of light in two, directing the signals out at right angles and then reflecting them back to a single focus. At the focal point, the two waves would interfere with each other to create bright and dark patterns, or fringes, whose width indicated the degree to which one beam had been accelerated relative to the other, or the degree to which the distance traveled by the two beams differed. Michelson and Morley realized that the movement of the apparatus in response to Earth's orbital motion would slightly alter the course of each light beam because the position of the reflecting mirrors and the focal point

Along with other scientists of the day, Americans Albert Michelson *(far left)* and Edward Morley believed that light, being a wavelike phenomenon, must travel through a medium, just as water carries waves in the ocean. In 1887, they used the device shown at right to try to detect the ether presumed to fill space. Hypothesizing that the speed of light beamed along the direction of Earth's motion through the ether should be different from the speed of light traveling perpendicular to that motion, they were disappointed to find no difference. Eighteen years later, Einstein explained the finding when he declared the idea of the ether unnecessary—inasmuch as the speed of light is absolute.

would shift slightly during the course of the experiment; thus the beam's outward journey might be longer or shorter than the return journey, and the total distance traveled by each beam could differ, producing a small amount of interference independent of any ether effect. But the scientists fully expected that by rotating the frame, they would observe peaks of interference when one beam moved against the ether wind while the other followed a path of less resistance. To their surprise and dismay, they found that no matter how they rotated the frame, they could not detect any significant change in the interference pattern.

The failure of the experiment to reveal what Michelson had expected was in fact an important accomplishment, for the result implied a great deal about the speed of light as well as about its hypothetical medium. According to Newton's principles, a beam of light from a source that is moving should acquire the velocity of the source in addition to the speed of the light itself. But the experiment done by Michelson and Morley suggested that the motion of the frame of reference—in this case, the motion imparted to the interferometer by Earth's travel in its orbit—had no effect on light's velocity.

This larger implication was not immediately apparent to Michelson, how-

ever, nor to others who spent years scrupulously refining the original experiment and verifying its results. But in the 1890s, a few bold theorists began to explore the significance of the finding. Among them was Dutchman Hendrik Antoon Lorentz, a precocious intellect who had begun studying for his doctorate at the University of Leiden at the age of seventeen and assumed the university's new chair in theoretical physics at the age of twenty-four. Fascinated and vexed by the Michelson-Morley experiment, Lorentz worked out what he called transformations, equations that could be used to relate time and distance measurements made in different frames of reference. This led Lorentz in 1904 to propound his principle of correlation, which stated mathematically what had been implicit in the Michelson-Morley results—that the speed of light and other electromagnetic radiation is independent of the frame of reference; in other words, the motion of the observer relative to the light source has no effect on the speed of light. If, for example, Earth were suddenly to leave its orbit and begin hurtling toward the Sun at ten thousand miles a second, the relative speed of the sunlight reaching Earth would not increase in the slightest. This paradoxical principle brought physics to the brink of a revolution that would be carried out by Albert Einstein.

THE INNER WORKINGS OF ATOMIC CLOCKS

In recent decades, a leap in timekeeping technology has enabled scientists to perform crucial experiments requiring superaccurate clockwork, including tests of Einstein's theory of relativity. Early in the 1900s, when Einstein proposed that time could be distorted by gravity and relative motion, clocks were too imprecise to register such effects. The invention of the quartz-crystal clock in 1929 raised timekeeping standards by substituting the steady, electronically induced oscillations of a crystal for the periodic motion of a pendulum or other friction-prone mechanisms. But even the best quartz clocks were subject to slight aberrations. To achieve greater accuracy for experimental purposes, scientists in the late 1930s began to explore ways of regulating quartz oscillators using the most reliable electromagnetic frequencies known: those generated by atomic particles.

This precision stems from the fact that each type of atom emits or absorbs a quantum of energy—a specific amount—when it shifts from a high-energy state to a low-energy state or vice versa. The energy is in the form of a photon, a carrier of electromagnetic force whose frequency is the same for all atoms of a given kind undergoing the same transition. By tapping into this atomic consistency, researchers were able to fine-tune the frequency of quartz-crystal oscillators, reducing their margins of error to less than a millionth of a second a year. Various atomic clocks were devised, including the hydrogen maser detailed here, which relies on amplified microwaves to regulate the crystal's frequency; a device employing a focused beam of cesium atoms is depicted on pages 110-111. Both were used in the 1970s in experiments that confirmed the theory of relativity. More recently, scientists have come up with an atomic timepiece of even greater accuracy—an ion trap *(pages 112-113)* that eliminates the tiny errors introduced by the motion of atoms by virtually immobilizing a charged particle with a laser before sampling its frequency.

As illustrated at near right, the hydrogen maser (an acronym for microwave amplification by stimulated emission of radiation) generates a signal by inducing a hydrogen atom in a high-energy state *(purple sphere)* to emit a photon *(purple squiggle)*, at which point the atom drops to a low-energy state *(blue sphere)*. Each photon emitted stimulates the release of other photons when it passes close enough to a high-energy atom to cause it to transition. The result is an amplified microwave signal with a frequency of 1,420,405,752 hertz (cycles per second), or 1.42 gigahertz, the frequency of each of the emitted photons. The strength of the microwave signal is only indirectly related to the accuracy of the clock; what matters is that the frequency be consistent. Picked up by a receiver at the top of the device, the signal is used to control the clock's quartz crystal. Without such tuning, the crystal would be stable to approximately a billionth of a second over an hour or so. With it, the stability of the quartz oscillator increases by a factor of 1,000 over as long as a day.

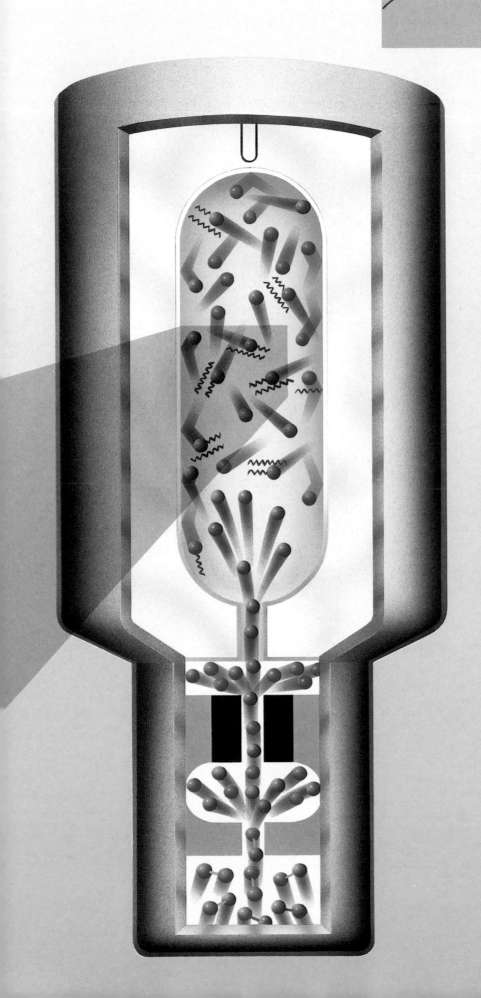

5 A small wire antenna projecting from the roof of the maser cavity picks up the 1.42 gigahertz signal. Through a series of electronic maneuvers, the maser uses this dependable signal—which completes 1.42 cycles every billionth of a second *(diagram above)*—as a standard to ensure that the quartz-crystal oscillator maintains a steady pace.

4 Inside the bulb, the high-energy atoms are bombarded by so-called blackbody radiation emanating from the walls of the surrounding cavity. (Such radiation is given off naturally by every object warmer than absolute zero.) The radiation includes a variety of radio-wave and microwave emissions, most of which have no effect. But when a high-energy hydrogen atom is approached by a photon at the atom's resonant frequency of 1.42 gigahertz, it responds by transitioning to a low-energy state and emitting a photon with the same frequency. The impinging photon and the one just generated will soon encounter more high-energy atoms and trigger more photon-emitting transitions, until the bulb is dominated by radiation at the 1.42 gigahertz frequency.

3 Next, the individual atoms pass between a pair of magnets *(black)*. Because an atom's magnetic properties vary according to its energy state, the low-energy atoms are deflected laterally by the magnets into the two wings of the chamber above, while the high-energy atoms are steered straight through into a bulb-shaped section of the maser.

2 The hydrogen molecules pass into a second chamber, where they are bombarded by an electrical current that shatters their bonds and splits them into atoms.

1 The bottom chamber of the maser is filled with molecules of diatomic hydrogen gas (pictured here as barbells); the molecules are composed randomly of high-energy atoms *(purple)* and low-energy ones *(blue)*.

Setting a New Standard

Hydrogen-maser clocks tend to lose their high degree of accuracy after a few days because the repeated collisions of hydrogen atoms against the interior wall of the bulb may cause shifts in frequency that will be picked up by the receiver and throw the quartz oscillator off stride. The atomic clock pictured here is more accurate because the cesium atoms involved in the process remain relatively unperturbed. Instead of interacting with a surface, the atoms are subjected to microwave radiation—generated by the quartz oscillator—that causes them to make a transition to another energy state. This transition, in turn, is one element of a feedback loop that regulates the oscillator.

The crystal oscillator's frequency is multiplied electronically, so that if the quartz is oscillating at the prescribed 5 megahertz (5 million hertz), the microwaves will radiate at cesium's resonant frequency of

The purpose of the cesium-beam clock, like the hydrogen maser, is to keep a quartz crystal (housed in the middle box below) oscillating at exactly 5 megahertz. Electronic circuitry in the box multiplies that frequency to produce microwaves with the resonant frequency of the cesium atom; a detector then collects feedback from the atoms that have passed through the microwaves to regulate the oscillator.

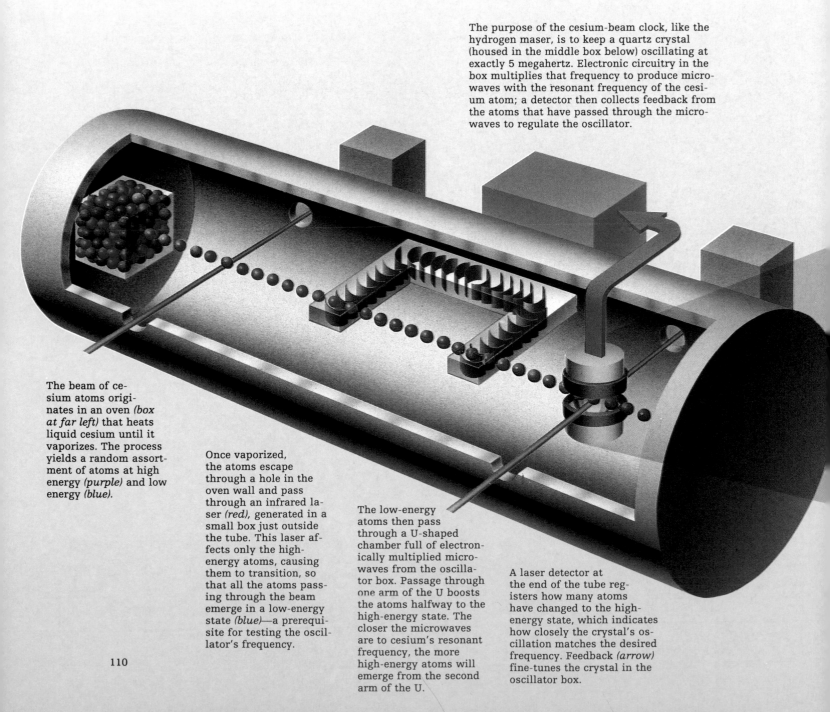

The beam of cesium atoms originates in an oven (box at far left) that heats liquid cesium until it vaporizes. The process yields a random assortment of atoms at high energy (purple) and low energy (blue).

Once vaporized, the atoms escape through a hole in the oven wall and pass through an infrared laser (red), generated in a small box just outside the tube. This laser affects only the high-energy atoms, causing them to transition, so that all the atoms passing through the beam emerge in a low-energy state (blue)—a prerequisite for testing the oscillator's frequency.

The low-energy atoms then pass through a U-shaped chamber full of electronically multiplied microwaves from the oscillator box. Passage through one arm of the U boosts the atoms halfway to the high-energy state. The closer the microwaves are to cesium's resonant frequency, the more high-energy atoms will emerge from the second arm of the U.

A laser detector at the end of the tube registers how many atoms have changed to the high-energy state, which indicates how closely the crystal's oscillation matches the desired frequency. Feedback (arrow) fine-tunes the crystal in the oscillator box.

110

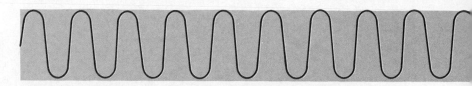

9.19 gigahertz, or 9.19 cycles every billionth of a second *(diagram, top right)*. This will induce low-energy cesium atoms beamed through the chamber to transition to a high-energy state. If, on the other hand, the crystal's oscillations begin to wander, then the microwaves in the chamber will stray from cesium's resonant frequency, and fewer atoms will transition.

The error will be spotted by a laser detector at one end of the tube. Low-energy atoms passing through the detector do not scatter photons in the laser beam as high-energy atoms would, causing signals to be sent to correct the oscillation frequency of the quartz crystal. The oscillator completes the feedback loop by tuning the microwaves to the precise resonant frequency of cesium. This increases the number of transitions to high-energy atoms, which in turn increases the incidence of scattering.

The device proved so accurate that in 1967, the second was redefined in terms of this transition: A second is officially the length of time it takes 9,192,631,770 cycles of microwave radiation to induce a transition between two energy states of the cesium atom.

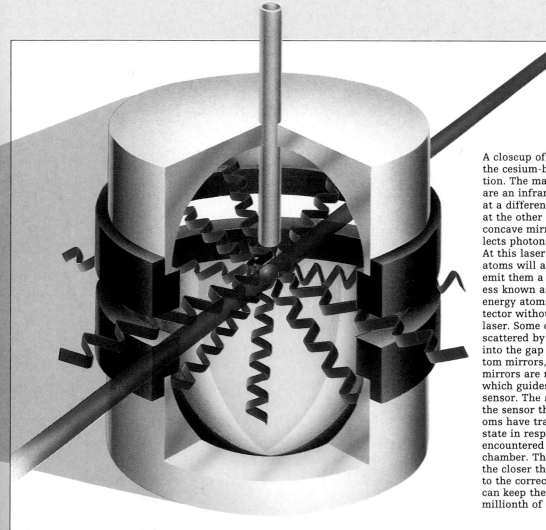

A closeup of the detector at the end of the cesium-beam tube details its operation. The main components of the device are an infrared laser *(red)*, which is set at a different frequency from the laser at the other end of the device; a pair of concave mirrors; and a pipe that collects photons reflected by the mirrors. At this laser's frequency, high-energy atoms will absorb laser photons and emit them a split second later—a process known as scattering—but low-energy atoms will pass through the detector without interacting with the laser. Some of the laser photons *(purple)* scattered by high-energy atoms escape into the gap between the top and bottom mirrors, but those that strike the mirrors are reflected into the pipe, which guides the light to an electronic sensor. The number of photons reaching the sensor thus reveals how many atoms have transitioned to a high-energy state in response to the frequency they encountered in the tube's microwave chamber. The more photons detected, the closer the crystal's oscillations are to the correct frequency. The feedback can keep the clock accurate to within a millionth of a second over three years.

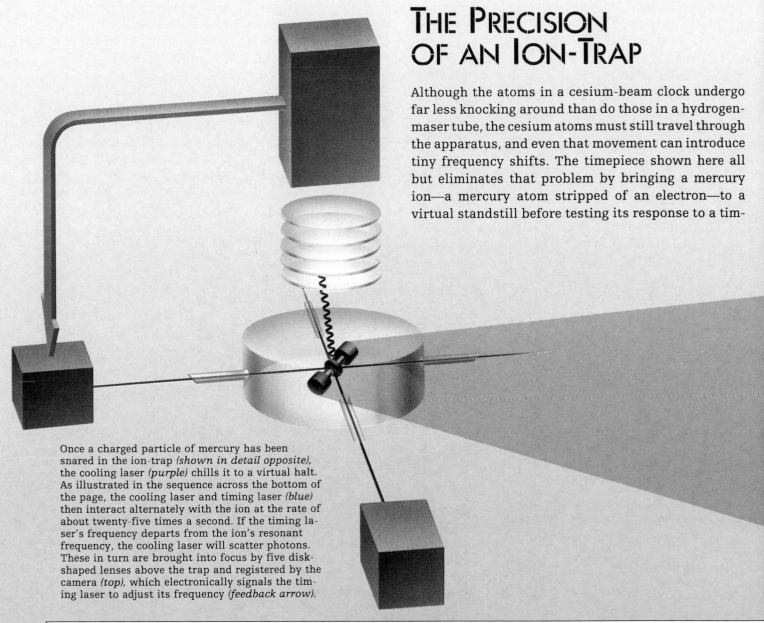

THE PRECISION OF AN ION-TRAP

Although the atoms in a cesium-beam clock undergo far less knocking around than do those in a hydrogen-maser tube, the cesium atoms must still travel through the apparatus, and even that movement can introduce tiny frequency shifts. The timepiece shown here all but eliminates that problem by bringing a mercury ion—a mercury atom stripped of an electron—to a virtual standstill before testing its response to a tim-

Once a charged particle of mercury has been snared in the ion-trap *(shown in detail opposite),* the cooling laser *(purple)* chills it to a virtual halt. As illustrated in the sequence across the bottom of the page, the cooling laser and timing laser *(blue)* then interact alternately with the ion at the rate of about twenty-five times a second. If the timing laser's frequency departs from the ion's resonant frequency, the cooling laser will scatter photons. These in turn are brought into focus by five disk-shaped lenses above the trap and registered by the camera *(top),* which electronically signals the timing laser to adjust its frequency *(feedback arrow).*

A low-energy ion *(light blue)* has been brought to a standstill by the cooling laser to eliminate motion that would cause a Doppler shift and skew the ion's response to the timing laser.

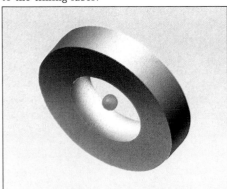

If the timing laser's frequency matches that of the ion, the ion will absorb a photon and jump to a high-energy state *(dark blue).* Before the photon can be emitted (scattered), the laser flicks off.

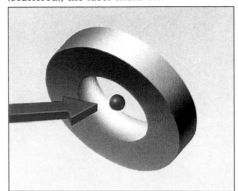

When the cooling laser flicks on, it cannot interact with the high-energy ion, so none of its photons are scattered. When this laser goes off, the ion emits the photon it absorbed from the timing laser.

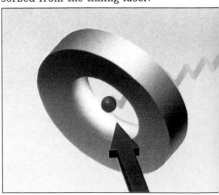

ing laser and using that response to tune an oscillator, which in turn runs a clock.

The instrument first traps the ion in an electromagnetic field. The ion is further restrained by a cooling laser *(purple beam)* set at a frequency just below one million gigahertz, mercury's resonant frequency. When the ion strays toward this laser, the ion's motion compresses the laser's light waves. The result is a Doppler shift that raises the light waves' apparent frequency to the ion's resonant frequency, causing the ion to scatter a photon—to absorb and emit it. In the process, the ion loses a bit of its momentum and cools. After thousands of such events, the ion is chilled to near absolute zero, leaving it practically immobile but still capable of scattering photons.

The slowed ion is then exposed alternately to the cooling laser and the timing laser *(blue beam)*. When the timing laser is attuned to the ion's resonant frequency, its effect interferes with that of the cooling laser, and no photon scattering is detected. If the timing laser strays from that frequency, the cooling laser scatters photons toward a camera; an electronic signal then adjusts the timing laser. The feedback is so precise that a clock regulated by the laser's frequency will err by no more than a billionth of a second over ten years.

The ion-trap consists of two cylindrical electrodes with a ring-shaped electrode between them. An electrical voltage applied to the ring electrode sets up an electromagnetic field that is weakest in the very center of the trap. The interaction between this field and the ion's own electrical charge forces the ion toward the weak spot in the center, confining its movement to a zone of about one cubic millimeter. The trap is positioned so that both lasers can aim their beams at the ion, but only photons scattered from the cooling laser will be focused by the lenses and picked up by the camera.

After scattering the timing laser photon, the ion returns to its ground state *(light blue)*. The photon has eluded the detectors, which are designed to pick up emissions triggered by the cooling laser only.

In this frame, the timing laser's frequency has drifted away from the ion's resonant frequency. As a result, the ion does not interact with the laser and remains in its original state.

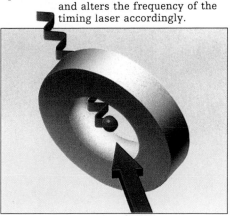

Unchanged by the timing laser, the ion now responds to the cooling laser by scattering a photon. The camera detects this photon and alters the frequency of the timing laser accordingly.

TRAVELING ON A LIGHT BEAM

Einstein's special theory of relativity, published in 1905, drew on the insights gained through an imaginative journey that he called a thought experiment. In this experiment, which he first conducted when he was a boy and which he took up time and again over the years, he envisioned himself riding a beam of light and tried to conceive of how events would unfold from this perspective. Since light transmitted information about the world to the senses, would someone traveling at light-speed be aware of any sequence of events? Would light itself become frozen in time?

As a physicist and mathematician, Einstein, of course, did not limit himself to thought experiments. He was concerned with effects that could be calculated. If his equations showed that the speed of light is both inflexible and independent of the motion of the observer, something had to give—and that something was the Newtonian concept of absolute space and time. In the case of Earth approaching the Sun at ten thousand miles a second, Newton's laws of motion predicted that the speed of the Sun's light as perceived from Earth should increase by the value of Earth's velocity, to a figure of some 196,000 miles per second. Yet Einstein was convinced, based on the work of Michelson, Lorentz, and others, that this was impossible; under no circumstances could the speed of light be increased. But if light did not accelerate, then space must shrink, an effect now called the Lorentz contraction, or time must stretch out, an effect known as time dilation.

In other words, the measurement of time depends on the relative motion of the observer. Einstein's equations showed that time dilation of a moving clock is negligible when the clock's speed is small compared to the speed of light. The effect increases with increased velocity, and becomes significant as the

If the speed of light is constant, said Albert Einstein *(left)* in 1905, then space and time are relative and there is no such thing as absolute simultaneity. Rather, the order of events can be experienced differently by observers in motion relative to one another. In diagrams depicting four-dimensional "space-time," mathematician Hermann Minkowski, one of Einstein's teachers *(far left)*, captured the new concept that space and time are interdependent.

clock's velocity approaches the speed of light. In such circumstances, a time-piece of any kind—wrist watch, pendulum, or water clock—would run slower, so that each tick or swing would take longer than that of timepieces in a stationary frame. To a hypothetical traveler carrying a clock at near light-speed, time would appear to pass normally, but each hour the clock advanced would encompass years in a fixed frame; such a traveler would return to his point of origin to find that he had remained relatively young while those left behind had aged considerably.

Among its many far-reaching implications, Einstein's special theory of relativity suggested that time is so closely related to space as to be one of its dimensions. After all, light-speed is a measure of distance traversed over time, and dilation amounts to the stretching of time to accommodate the greater distance covered by one body relative to another. The first to come up with a way of expressing this interrelationship graphically was a German mathematician and former teacher of Einstein's named Hermann Minkowski, who concluded in 1908 after studying his prize pupil's special theory that time should be considered graphically as the fourth dimension of space, along with length (longitude), width (latitude), and height (altitude).

Minkowski's concept of space-time, readily adopted by Einstein, was based on an established method of perspective drawing that allowed one to represent spatial events such as the orbit of a planet around the Sun in three dimensions by projecting three axes: x, y, and z. To represent the same event in time, however, one would have to add a fourth line, designated t, something no drawing can cope with. So Minkowski omitted one dimension—height—and substituted time. A momentary occurrence would appear on such a graph as a "world point," as Minkowski put it, while any event that persisted would yield what he called a "world line." In the case of a planet's orbit, then, Minkowski's scheme would require that the orbit be depicted as flat, an acceptable compromise, since most planets travel within or near the Sun's equatorial plane. That would leave room to represent the shape of the orbit in two dimensions, with time as the third dimension.

TIME AND GRAVITY

Einstein's special theory of relativity applied only to bodies that were moving at constant speeds. But in the years following the publication of that theory, he began to look at the more complex issue of accelerated motion. Along the way, he happened upon a crucial insight, inspired in part by a seemingly trivial event—an accident that befell a man in his neighborhood in Berlin, who survived a tumble from the roof of a building and commented afterward with apparent surprise that he had not felt the tug of gravity on the way down. That simple observation inspired what Einstein called "the happiest thought of my life."

He realized that "if one considers an observer in free fall, for example from the roof of a house, there exists for him during his fall no gravitational field—at least in his immediate vicinity." The force of gravity would not become apparent until the free fall was decelerated, either abruptly by an encounter with the ground, or more gently if one happened to be equipped with a parachute. Conversely, someone who was floating weightlessly in space within a container (a situation that Einstein imagined in another one of his thought experiments) would feel the effect of gravity if the container were to accelerate. Such speculation led to what Einstein termed the principle of equivalence, which states that the forces of acceleration and gravity are identical in effect.

Einstein showed that an accelerated observer quickly builds up a finite velocity, which affects his clocks and measuring rods when compared to those belonging to an observer who is not accelerating. Then, according to the principle of equivalence, a gravitational field must also have an effect on the rate at which a clock ticks. For example, a clock placed near the surface of a star will tick a little slower than one farther away from the star. Here, as with motion at a constant speed, the result would be a slowing of the clock, with insignificant effect for a star of small mass, and a mounting impact for an extremely massive star.

This revelation laid the theoretical foundation for Einstein's general theory of relativity, published in 1915, which dealt with the warping of space-time by extremes of gravity and acceleration. To science-fiction enthusiasts, Einstein's predictions suggested fascinating scenarios, many of them involving spacefarers and their stay-at-home twins. In a typical case, a spacefarer might leave Earth for a ten-year mission and return to discover that his twin had aged forty years. That outcome might indeed be possible if the spacefarer's craft could reach fantastically high speeds, much closer to the speed of light than can be achieved by booster rockets today. But such dramatic scenarios tend to overlook the trade-offs between the time-dilating effects of high relative speed and any periods spent traveling at more realistic speeds or, more pertinent, in a gravity-free environment—in which case the Earthbound twin, under gravity's constant effect, could age more slowly. In some circumstances, the spacefarer might return to find his twin a few seconds younger, hardly a stirring conclusion for a science-fiction story, perhaps, but one that would be more faithful to the intricacies of relativity.

TESTING THE THEORIES

For decades, such hypothetical cases were the only use Einstein's predictions could be put to. Before his findings could be applied, they had to be confirmed, and that required superaccurate timepieces, since the largest relativistic effects that could be reproduced experimentally would involve time discrepancies far too small to be detected with clocks available at the time. With the development in the late 1940s and 1950s of atomic clocks and other devices for gauging the precise frequencies given off by excited atoms or molecules, researchers at last had instruments that were sensitive enough to test the theories *(pages 108-109)*. In 1960, a group of physicists at Harvard University led by Robert V. Pound and Glen A. Rebka transmitted the frequency given

NAVIGATING BY RELATIVITY'S RULES

Satellites equipped with atomic clocks that take into account the time-distorting effects of relativity have revolutionized navigation. Through a sophisticated application of a simple formula—elapsed time multiplied by speed equals distance—navigators can fix their whereabouts by computing how long it takes radio signals from such satellites to reach their craft. A receiver on the craft picks up the signal, which is coded to reveal the exact time the satellite dispatched it; the same instrument then multiplies the elapsed interval by the speed of radio waves (186,282 miles per second) to determine the distance to the satellite. By comparing readings from three or more satellites whose positions are known, navigators can zero in on their own location.

For the system to work, measurements of the time it takes signals to dart from satellite to receiver must be precise. An error of just one-thousandth of a second means a discrepancy of 186 miles. To maintain accuracy, each satellite carries four atomic clocks—three backups in case of malfunctions. And although most receivers rely on a less-accurate quartz clock, navigators can compensate for the margin of error by taking readings from additional satellites (page 118).

The system must account as well for two time-warping factors identified by Einstein—gravity and relative motion. As shown at right, the farther a clock is from Earth's center of gravity, the more quickly it ticks. On the other hand, the faster an object moves relative to another, the slower its clock.

Many navigation satellites circle the planet twice a day so that their positions can be monitored every twelve hours from ground stations. By itself, this velocity relative to Earth causes satellite time to lag behind ground time, although the slowdown is less than the speed-up resulting from the gravity factor. Engineers compensate for the combined effects by programming the satellite to subtract the net gain in time before sending its signals. The readings from each satellite are then compared to the time indicated by a master clock on the ground and adjustments made if necessary to ensure that all the satellites remain synchronized to a second.

A clock on a navigation satellite thousands of miles out in space will run faster than a clock on the ground because it is farther from Earth's center of gravity. The actual difference is only millionths of a second per day; for purposes of illustration, the clock here shows a gain of twenty-four minutes.

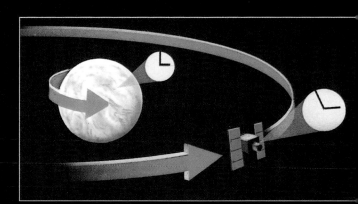

In a separate relativistic phenomenon, the satellite's faster orbital motion (large arrow) relative to Earth's rotation (small arrow) results in the satellite clock lagging behind the one on the ground. Expressed in proportion to the gravitational example above, the orbiting clock has lost about four minutes.

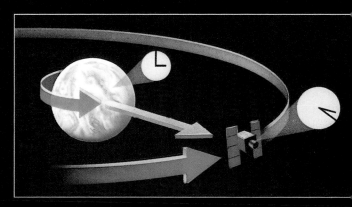

The combined effects of gravity and motion produce a net gain (shown here as about twenty minutes) for the satellite clock. The actual gain over the course of a day, although much smaller, would skew distance calculations by 7.2 miles. By accounting for the gain, the satellite allows for pinpoint navigation.

Solving a Puzzle in Three Dimensions

Theoretically, navigators can find their bearings by taking one satellite distance reading for each dimension to be plotted, a process called triangulation. In the case of an orbiting space shuttle, for example, three measurements—for altitude, latitude, and longitude—should suffice. Each measurement may be thought of as a two-dimensional plane, and the point where the planes intersect would be the shuttle's location. In practice, the results are never that neat.

Even after correcting for relativistic effects, the timing will remain imperfect and the shuttle's position

only approximate unless it too is equipped with an atomic clock that is synchronized with a master ground clock. A less costly solution is to incorporate a fourth satellite reading, as shown here.

Because the shuttle clock is not perfectly in sync with the satellite clocks, each reading will locate the shuttle not in a flat plane but in a zone with a thickness that corresponds to the margin of error for the shuttle clock. The intersection of three such zones (shown below as disks) will be a small region within which the shuttle might be located. The fourth reading serves to ascertain the shuttle's synchronization error—only one value for that error will bring the fourth plane into contact with the intersection of the other three. Once that error is found, the other measurements become precise enough to fix the shuttle's position.

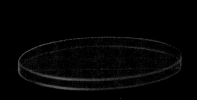

A satellite orbiting above the shuttle reveals the craft's altitude. Because the shuttle's clock is not synchronized with the satellite's, the measurement is shown as a disk whose thickness indicates a range of possible locations.

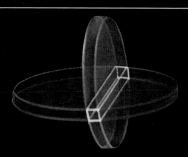

By including data from a satellite on the horizon, the range for the shuttle is narrowed to the cubic intersection of two disks (outlined in white). If the timing were perfect, the disks would be flat and intersect along a line.

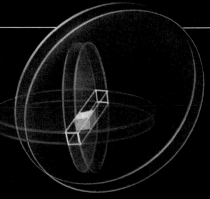

A third measurement restricts the location to a small cube (solid white). A fourth reading (right) eliminates the margin of error and yields a single point.

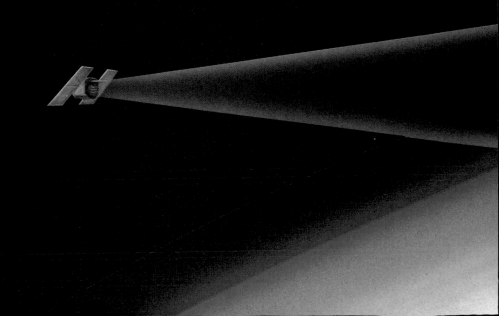

To be effective worldwide, an orbiting navigation system must keep several satellites within radio range of all points at all times. The United States Department of Defense is scheduled to complete such an array in 1992 when it launches the last of the twenty-four satellites in its Global Positioning System, or GPS. With four satellites in each of six orbits *(above),* GPS will provide exhaustive coverage; at any time or place, a receiver will be able to gather readings from at least six satellites—more than enough to meet any need. U.S. military receivers will be equipped to decode satellite signals down to the last digit, allowing positions to be resolved within approximately twenty yards. For security reasons, civilian receivers will be slightly less sophisticated, restricting accuracy to around a hundred yards.

off by a radioactive form of iron, Fe^{57}, to a receiver whose position in relation to the source was higher or lower by seventy-four feet. The physicists found that the resulting difference in gravity, although minuscule, nonetheless shifted the frequency of the signal—or changed its timing—by an amount that came within ten percent of the value predicted by Einstein. An improved version of the experiment performed in 1965 brought the result to within one percent of Einstein's figure.

Then in 1971, scientists at Washington University in St. Louis and the U.S. Naval Observatory flew four cesium-beam atomic clocks around the world, once circling the globe west-to-east and then again traveling east-to-west, to test the combined time effects of gravity and relative speed. Before and after each trip, the clocks were compared with clocks at the Naval Observatory.

VALIDATION

In terms of gravity alone, the airborne clocks were expected to run faster than the ground clocks. But in the case of the eastward-flying plane, that effect would be offset because the jet would travel faster than the eastward-moving Earth and thus would move faster than the ground clock; this velocity factor would cause the clock to lose slightly more time than it gained due to the gravity differential. By contrast, the westward-flying jet would move against the direction of Earth's rotation and would recede relative to the advancing ground clock; this would cause the westward jet's clock to run that much faster. The results of the experiment bore out these predictions. The expected gain for the westward journey was 275 nanoseconds (billionths of a second), and the actual gain recorded was 273 nanoseconds. For the eastward journey, the observed loss of 59 nanoseconds compared to a predicted loss of 40 nanoseconds. The combined figures were close enough to the mark to validate Einstein's theories.

Further confirmation came from a stratospheric test conducted by Robert Vessot and Martin Levine of the Smithsonian Astrophysical Laboratory at Harvard University in collaboration with the National Aeronautics and Space Administration. On June 18, 1976, the team lofted a hydrogen-maser clock on a small scout rocket into suborbital flight. In its two-hour excursion into an area of low gravity, the clock registered the anticipated speed-up before falling into the ocean east of Bermuda. The frequency shift amounted to about 1 hertz in 1.42 gigahertz, with a margin of error of just .007 percent.

Today, these minute but demonstrable effects play a significant role in time-sensitive satellite navigation systems such as the Global Positioning System *(page 117)*. Atomic clocks aboard satellites in high Earth orbit are programmed to compensate for the relativistic effects of gravity and motion, allowing users at lower altitudes to zero in on their own position with unprecedented accuracy.

While research scientists testing out the principles of relativity were reaching new heights of precision, theoretical physicists following in Einstein's footsteps were entering realms filled with uncertainty. Among the ticklish questions pondered by the theorists was whether the universe described by Einstein is closed or open-ended. In one model, the universe has finite dimensions but at the same time lacks boundaries—an idea that can be visualized by picturing a person walking around a featureless globe. This world would have no edge or ending place, and yet its total area would be limited. The problem with this analogy is that the real "globe" as defined by Einstein has four dimensions, one of which is time. And once time is linked to space, the present shape of the universe becomes much harder to picture. Furthermore, the question arises as to how the shape of the universe has changed with time.

Astronomers analyzing the spectra of starlight from distant galaxies have concluded that the frequency of that light is shifted from its true value in a way that can be explained only if the galaxies are retreating from Earth at a constant rate. Evidently, the universe is steadily expanding, which suggests in turn that the cosmos began in the distant past as an infinitely dense point without dimension. That point could be considered the beginning of time, a state of affairs cosmologist Steven Weinberg has likened to absolute zero. In this sense, the Big Bang—the convulsion that launched the expansion of the universe—set time as well as matter and space into motion.

In 1976, Martin Levine *(above left)* and Robert Vessot of the Smithsonian Astrophysical Observatory sent a hydrogen-maser clock on a low-gravity suborbital flight to test Einstein's predictions about the influence of motion and gravity on time. As the device accelerated at launch, the oscillations of its hydrogen atoms seemed to slow; midflight, weaker gravity caused an apparent speed-up.

The realization that the flow of time is linked to the growth of the universe has perplexing implications. Some cosmologists predict that the universe will eventually exhaust the momentum of expansion and begin to contract. If so, would time also change direction, and would existing sequences of cause and effect be reversed in the process?

Many physical processes involve periodic movements that would operate equally well in reverse. The orbit of a planet, the swinging of a pendulum, the exchanges of energy within the atom—all of these would shift direction if time flowed backward but would still obey the same physical principles. But other physical processes could not be reversed without violating the rules as we know them. Several scientists have raised fundamental objections to the idea that nature could shift gears, as it were, and run backward as neatly as it moves forward. In the early part of the century, the Cambridge University astronomer Sir Arthur Eddington invoked the phrase "the arrow of time" to describe the powerful sense of direction he perceived in the natural world, with its sequences of growth and decline. Later, physicist Roger Penrose of Oxford University expanded on Eddington's concept, focusing on the universal pattern of entropy—nature's tendency to increasing disorder—as the most obvious of time's arrows: Structures eventually collapse, fires burn out, heat dissipates, radioactive elements decay, organisms expire and molder. To imagine a contracting universe following the script of the expanding universe in reverse is to deny entropy in seemingly absurd ways. Thus bodies would reconstitute themselves from bones and dust and emerge from the grave to grow younger and stronger by the day, or a giant asteroid would spring from the ashes of a vast crater on a ruined planet and fly off into space, leaving a green world teeming with life in its wake.

ODDS AGAINST

Stephen Hawking, a mathematician and cosmologist at Cambridge University, has looked at the possibility of reversing entropy in terms of quantum theory, which states that an observer cannot precisely measure the location and the velocity of a subatomic particle simultaneously and thus has a limited ability to predict the subsequent behavior of such a particle. All that the observer can do in these circumstances is to evaluate the statistical likelihood that particles of that description will follow a particular course amid a range of possibilities. A scientist trying to imagine how a relatively simple instance of entropy—such as a teacup falling from a table and breaking on the kitchen floor—would play in reverse is faced with a similar problem, Hawking observes. Since gravity too might be reversed, one could not simply dismiss the possibility that a piece of broken china would leap up from the floor onto the table. But for a cup to form, each piece would have to follow the exact path by chance. The odds that all of them would come together at the right time and place are fantastically long. Hawking concludes that a contracting universe would tend to "a very disordered state," with little chance of any order emerging from the chaos as one would expect if the principle of entropy applied in reverse.

Aside from its implications for the future, quantum theory has an immediate bearing on the meaning of time and the structure of events. The principle of uncertainty threatens to overturn the very idea of cause and effect and a longstanding scientific belief that the laws of nature can be measured objectively and quantified precisely—a conviction summed up in grand fashion by the eighteenth-century French mathematician Pierre-Simon Laplace, who wrote that the ultimate goal of science was to "condense into a single formula the movement of the greatest bodies of the universe and that of the lightest atom." To an intellect that could grasp the formula, Laplace insisted, "nothing would be uncertain," and the future and the past would be as intelligible as the present. Even Einstein, whose theories did much to raise the level of uncertainty in physics, shared this belief that the laws of nature could be reduced to coherent and verifiable terms. But the search for a grand formula that would unify the fundamental forces and explain everything is clouded by the quantum axiom that the very process of observing subatomic events can affect their outcome. For example, firing other particles at atoms, as researchers do in accelerators in order to learn about particle behavior, inevitably changes the motion of the targets. The observational tools are so powerful that they alter the subject under study.

More disconcerting is the inference drawn by quantum physicists that on the very smallest scale—beyond any present means of detection—ghostly fragments called virtual particles are popping into and out of existence. Indeed, physicist John Wheeler has suggested that this evanescent "quantum foam," as he calls it, may exist in a strange state where there is no before or after, no cause or effect. If scientists were somehow to devise an instrument that could penetrate this realm without altering it beyond recognition, they might find it impossible to apply their conventional standards of comparison and measurement. For this reason, Wheeler says, physicists may simply have to discard the traditional standard of time when studying such basic phenomena. At its earliest, deepest, and densest levels, nature may be without sequence or structure.

THE RIDDLE OF TIME'S PASSAGE

One of the inescapable facts of life is that time flows in only one direction: from past to future. Even as today fades into yesterday, tomorrow is becoming today, and—except in the movies—the reel can never run in reverse. Despite its grounding in everyday experience, time's one-directional mode seems to conflict with many of the laws governing the behavior of matter in the universe. The mathematical logic of these laws is such that they are time-symmetric. That is, time's direction is irrelevant, and there is no reason, in theory, why time should not readily run backward.

That it simply does not suggests that the universe somehow distinguishes between the two directions, and, furthermore, that it overwhelmingly prefers time to move forward. In trying to discover why this should be so, physicists have identified a few arrows of time, as they are known—processes and phenomena that appear to exemplify time's one-way passage.

For a process to be classified as an arrow of time, it must work either differently or not at all if time were reversed. Since scientists cannot actually reverse time, of course, they resort to so-called thought experiments, like those illustrated on the following pages, in which the process can be run backward mentally—or by computer—and the result analyzed. Following through on the movie analogy, a process is mentally "filmed" and then viewed both forward and backward. In most instances, the process is time-symmetric and the film makes sense both ways. On rare occasions, however, the film makes sense only when run forward, or plays sensibly but differently when run backward. Such a time-asymmetric process may offer a clue to the riddle of why time flows only one way.

Among the many examples of observable physical processes that make no distinction as to the direction of time are the laws governing planetary motion, first elucidated by Johannes Kepler in 1620 and incorporated into Newton's theory of universal gravitation in 1687. In the illustration at left below, time is moving forward, and the planets in a hypothetical solar system travel in clockwise orbits around their sun. At right, time flows backward; the planets reverse their orbits and travel counterclockwise. Even though everything is moving in the opposite direction, all the laws of planetary motion remain intact: A film of the planets in motion would run equally well in either direction. The laws of celestial mechanics—and, it turns out, all of Newtonian mechanics—fail to shed light on time's inexorable forward march.

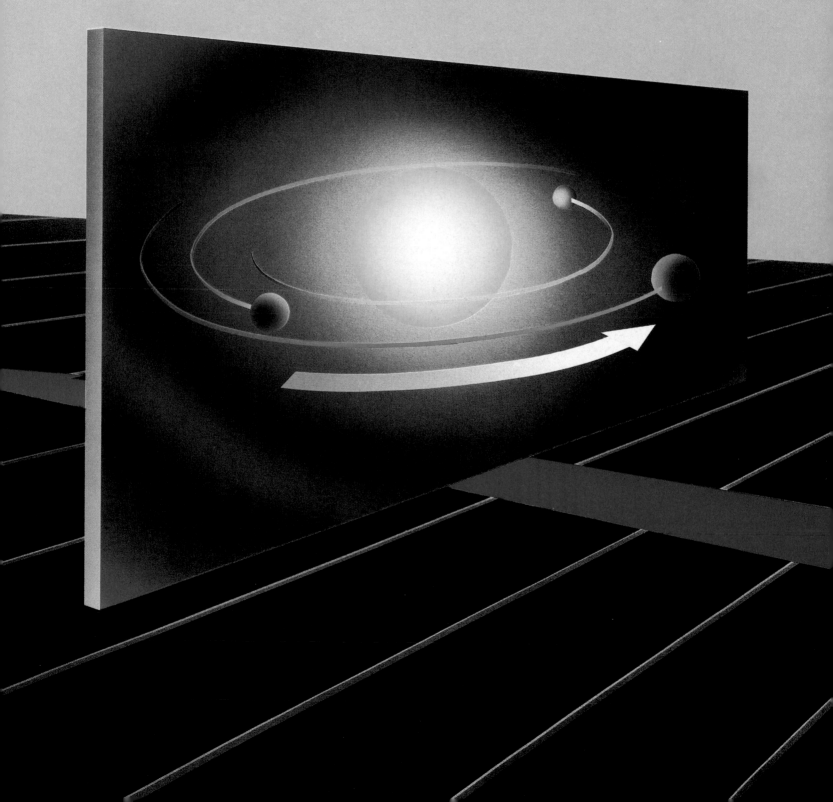

THE POWER OF ENTROPY

One clue to the puzzle may lie in what is perhaps the most comprehensive physical law in the universe: the second law of thermodynamics, which declares that entropy—the amount of disorder in a closed system—can never decrease. This law, which governs all processes everywhere, explains, for example, why broken objects never spontaneously reassemble themselves.

The concept of entropy is illustrated below, using the example of a recently opened bottle of liquid in a

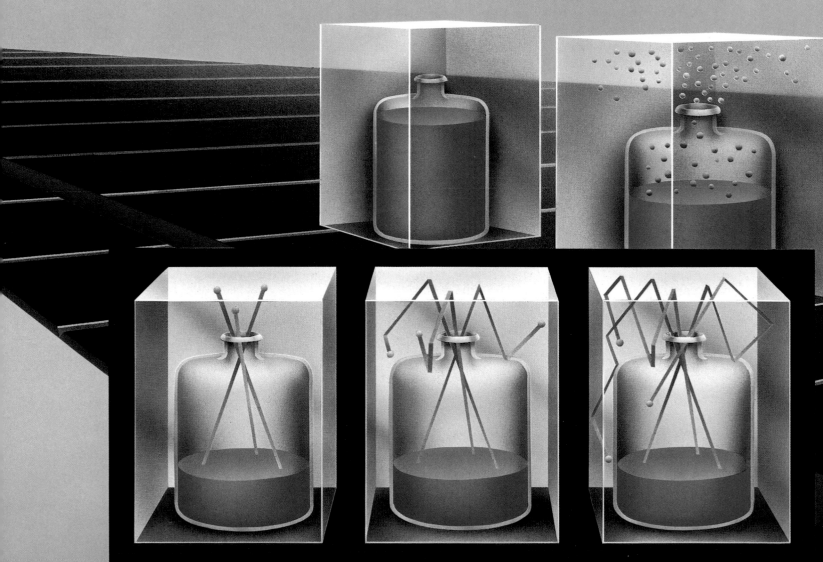

According to the second law of thermodynamics, any given gas molecule escaping from a bottle could in theory bounce back into it *(above)*, but the probability that all the molecules will spontaneously return is vanishingly small.

larger sealed container. As time passes, the molecules of the liquid evaporate, escaping from the bottle, bouncing off its walls, and filling the surrounding container. When all the liquid has evaporated, the gas molecules will have diffused into the space inside and outside the bottle—a less ordered state than before.

If time ran backward, the hypothetical film would show gas molecules bouncing off the walls of the outer container and back into the bottle, where they would recondense. However—again according to the second law of thermodynamics—the probabilistic behavior of large numbers of molecules dictates that such a scenario cannot occur. That is, when a molecule is considered individually, there is no physical law to prevent it from finding its way back into the bottle *(inset)*. But when the behavior of the molecules is examined statistically, the chance of all of them returning to the bottle is virtually nil.

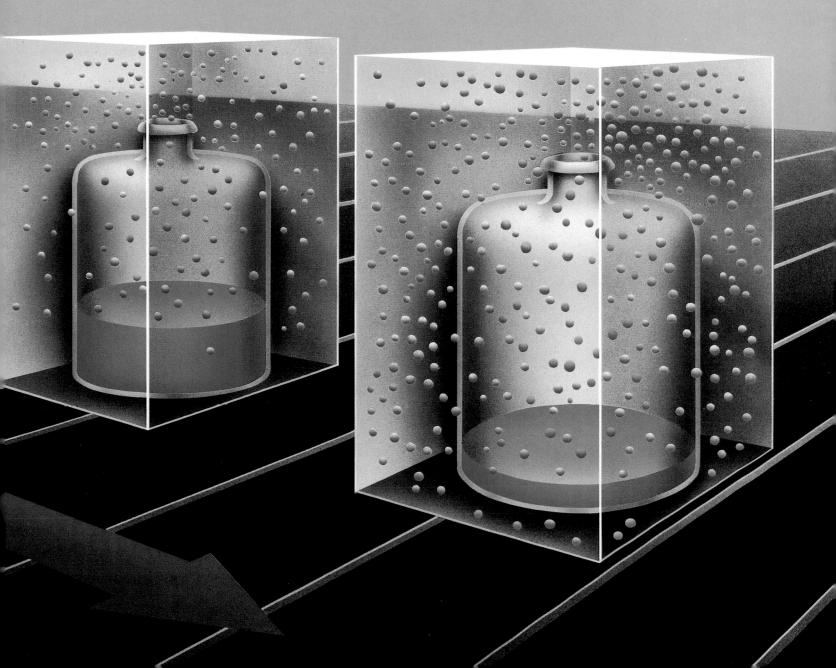

The Cosmological Arrow

Building on meticulous observations that most galaxies are racing away from one another—in some cases at rates approaching the speed of light—astrophysicists in the late 1920s and early 1930s concluded not only that the universe is expanding but that it was once an infinitesimally small seed. Fifteen or 20 billion years ago, most scientists believe, that primordial seed exploded in the so-called Big Bang that gave birth to all matter and to space-time itself. Time's direction, physicists say, was inherent in this explosion. Simply put, if the universe is getting larger, as depicted schematically below, time must be moving forward.

In an expanding universe, the future holds different

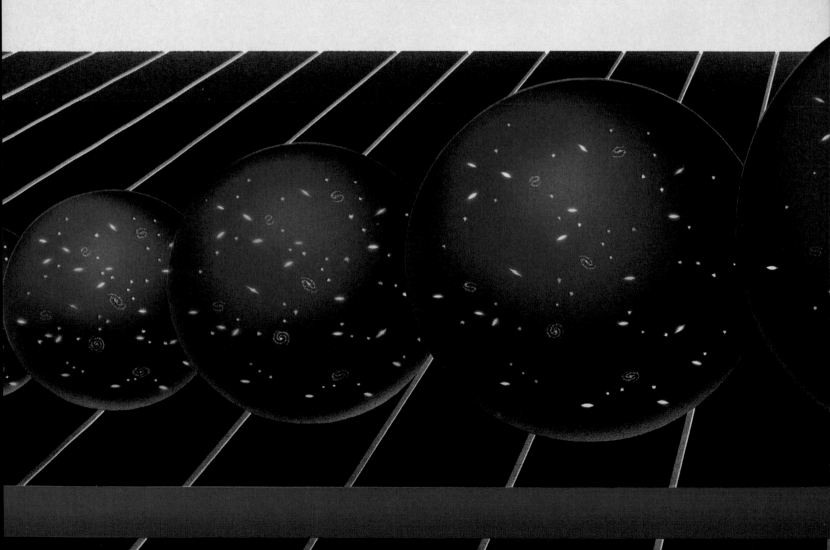

possibilities, depending on the density of the mass the cosmos contains. If it has too little mass density—and therefore gravity—to counteract the expansive force, the universe will continue to grow forever. The expansion will always define the forward direction of time.

If the universe has enough mass density to overcome the expansion and begin contracting upon itself, however, physicists disagree about the outcome. The first possibility is that the cosmological arrow of time will be redefined: Time will continue to move forward even as the universe contracts. The second option is that at the moment the contraction begins, time will reverse and the entire cosmos will replay its history backwards. To the inhabitants of this shrinking universe, however—whose very brain processes would go into reverse—time would still appear to be moving forward so that the cosmological arrow would be redefined in either case.

The Kaon's Break with Time Symmetry

Perhaps the strangest arrow of time is associated with one of the most obscure physical processes ever examined: the decay of a subatomic particle called the neutral kaon *(pages 44-45)*. Physicists studying the decay of kaons and other subatomic particles often resort to metaphors, such as the one illustrated above, which uses lenses to represent three laws of physics that are either observed or violated in the course of subatomic reactions.

According to the law of charge conjugation, for example, there should be no observable difference in a subatomic process if all the participating particles were replaced with their antimatter counterparts—

particles that are equal in mass but opposite in charge. According to the law of parity, an observer should not be able to distinguish between a given subatomic reaction and that reaction's mirror image. Finally, the law of time symmetry posits that a subatomic process should work the same way whether it operates forward or backward.

In the language of the lens metaphor, the symmetry that is represented by a given lens holds if a projected image (in this case, a circle) passes through the lens undistorted. A distorted image implies that the symmetry law is violated.

The diagram above shows how the charge, parity, and time symmetry laws work in all observed subatomic decay processes except that of the neutral kaon. Sometimes charge symmetry is not maintained, as seen in the distorted circle that emerges from the first lens *(far left)*. However, whenever the law of

charge symmetry is violated, there is an accompanying, and perfectly compensating, violation of the law of parity symmetry *(middle)*. Thus, the metaphoric result is the restoration of the circle as it passes through the second lens. The undistorted passage of the circle through the time lens represents the time symmetry of all observed nonkaon reactions.

The singular case of neutral kaon decay is illustrated in the diagram above. Physicists discovered that although violations still occurred in charge and parity symmetry, the second violation did not always completely compensate for the distortion caused by the first. However—and here was the puzzler—the result of the experiments was nevertheless the same as that for nonkaon reactions; in other words, it was analogous to the circle emerging from the third lens completely undistorted.

The only possible conclusion is that, in the case of

kaon decay, and unlike all other processes, a corrective distortion must have occurred in the third lens, representing time symmetry. And if time symmetry is violated, then the neutral kaon must decay in a different manner in a time-reversed universe.

Although physicists have no idea exactly how—or, for that matter, why—the decay would proceed differently, they believe that without this time asymmetry the present universe would not exist. According to classical cosmological theory, shortly after the Big Bang, the equal amounts of matter and antimatter that should have been created ought to have annihilated each other completely. That the cosmos exists, it now appears, is due to a flaw in the time symmetry of a hypothetical primordial particle called the Higgs boson. As a result, a slight excess of matter was created that was able to outlast its antimatter nemesis, going on to generate everything observed in the universe.

133

String theory: a theory that subatomic particles arise from tiny, one-dimensional strings, and that their properties are determined by the arrangement and vibration of the strings.

Strong force: the force that binds quarks together into composite particles and holds protons and neutrons together to form atomic nuclei.

Superstring theory: a supersymmetrical version of string theory.

Supersymmetry: a theory in particle physics proposing that every type of fermion or boson has a matching partner particle differing from it only in spin.

Symmetry: used by physicists to denote quantities that remain unchanged despite transformations in the system affecting them.

Symmetry group: a mathematical grouping of particles possessing a common property that unites its members and demonstrates a symmetry.

Theory of everything (TOE): a comprehensive theory that will explain all known physical phenomena, including gravity, in quantum terms.

Topology: the study of geometric forms that focuses on the relationship between surfaces, and on discontinuities (or holes) within a given surface.

Twistor: a theoretical entity that operates in an eight-dimensional complex space underlying space-time.

Uncertainty principle: the understanding that uncertain values are inevitable at the subatomic level since measuring techniques disrupt the particles being measured. For example, a particle's precise position and its momentum can never be known at the same time.

Velocity: the speed and direction of motion.

Virtual particles: extremely short-lived particles created out of nothingness, as permitted by the uncertainty principle. Although they exist too briefly to be directly observed, the effects of their existence may be detected.

Virtual photons: virtual particles that transmit the electromagnetic force. For example, the exchange of virtual photons is responsible for the electric repulsive force between two electrons.

Wavelength: the distance from crest to crest or trough to trough of an electromagnetic or other wave. Wavelengths are related to frequency: the longer the wavelength, the lower the frequency.

Weak force: a very short range force responsible for radioactive decay and other subatomic reactions.

World line: the one-dimensional path through space and time of a traditional point particle.

World sheet: the two-dimensional surface occupied by a string in its space-time history. Any point on a world sheet can be described by two numbers: one specifying the position of the point on the string, and the other the time.

Wormhole: a hypothetical distortion in the fabric of space-time produced by quantum fluctuations and linking widely separated black holes.

X Higgs boson: theoretically, a massive boson capable of releasing the strong force from the previously unified electronuclear force. X Higgs bosons would have existed only during a brief period beginning 10^{-35} second after the Big Bang; their decay would have produced an excess of matter over antimatter particles that is still evident.

BIBLIOGRAPHY

Books

Abbott, Edwin A. *Flatland: A Romance of Many Dimensions*. New York: Dover Publications, 1952.

Alexander, H. G. (ed.). *The Leibniz-Clarke Correspondence*. Manchester, England: Manchester University Press, 1956.

Andrews, William, and Seth Atwood. *The Time Museum: An Introduction*. Rockford, Ill.: The Time Museum, 1983.

Asimov, Isaac. *Asimov's Biographical Encyclopedia of Science and Technology* (2d rev. ed.). New York: Doubleday, 1982.

Bronowski, J. *The Ascent of Man*. Boston: Little, Brown, 1973.

Bruton, Eric. *Dictionary of Clocks and Watches*. New York: Archer House, 1963.

Butler, S. T., and H. Messel (eds.). *Time*. Oxford, England: Pergamon Press, 1965.

Calder, Nigel. *Einstein's Universe*. New York: Penguin Books, 1979.

Čapek, Milič (ed.). *The Concepts of Space and Time: Their Structure and Their Development*. Dordrecht, Holland: D. Reidel, 1976.

Close, Frank, Michael Marten, and Christine Sutton. *The Particle Explosion*. New York: Oxford University Press, 1987.

Cohen, I. Bernard:
The Birth of a New Physics. New York: W. Norton, 1985.

Revolution in Science. Cambridge, Mass.: The Belknap Press of Harvard University Press, 1985.

Coleman, Lesley. *A Book of Time*. New York: Thomas Nelson, 1971.

Considine, Douglas M., Jr., and Glenn D. Considine (eds.). *Van Nostrand's Scientific Encyclopedia* (7th ed.). New York: Van Nostrand Reinhold, 1989.

The Cosmos (Voyage Through the Universe series). Alexandria, Va.: Time-Life Books, 1988.

Cowan, Harrison J. *Time and Its Measurement: From the Stone Age to the Nuclear Age*. Cleveland: World Publishing, 1958.

Crease, Robert P., and Charles C. Mann. *The Second Creation: Makers of the Revolution in Twentieth-Century Physics*. New York: MacMillan, 1986.

Davies, Paul C. W. *The Forces of Nature*. Cambridge, England: Cambridge University Press, 1979.

Davies, Paul C. W. (ed.). *The New Physics*. Cambridge, England: Cambridge University Press, 1989.

Davies, Paul C. W., and Julian Brown (eds.). *Superstrings: A Theory of Everything?* Cambridge, England: Cambridge University Press, 1988.

Duncan, Ronald, and Miranda Weston-Smith (eds.). *The Encyclopaedia of Ignorance: Everything You Ever Wanted to Know about the Unknown*. Oxford, England: Pergamon Press, 1977.

Everitt, C. W. F. *James Clerk Maxwell: Physicist and Natural Philosopher*. New York: Charles Scribner's

Sons, 1975.

Fagg, Lawrence W. *Two Faces of Time*. Wheaton, Ill.: The Theosophical Publishing House, 1985.

Ferris, Timothy. *Coming of Age in the Milky Way*. New York: Doubleday, 1988.

Feynman, Richard P. *QED: The Strange Theory of Light and Matter*. Princeton: Princeton University Press, 1985.

Feynman, Richard P., Robert B. Leighton, and Matthew Sands. *The Feynman Lectures on Physics*. Reading, Mass.: Addison-Wesley, 1965.

Flood, Raymond, and Michael Lockwood (eds.). *The Nature of Time*. Oxford, England: Basil Blackwell, 1986.

Foster, B., and P. H. Fowler (eds.). *40 Years of Particle Physics*. Bristol, England: Adam Hilger, 1988.

Fritzsch, Harald. *Quarks: The Stuff of Matter*. New York: Basic Books, 1983.

Gardner, Martin. *The New Ambidextrous Universe* (3d rev. ed.). New York: W. H. Freeman, 1990.

Gillispie, Charles Coulston (ed.). *Dictionary of Scientific Biography* (Vol. 11). New York: Charles Scribner's Sons, 1980.

Goudsmit, Samuel A., and Robert Claiborne. *Time*. New York: Time, 1966.

Gray, H. J., and Alan Isaacs (eds.). *A New Dictionary of Physics*. London: Longman, 1975.

Green, Michael B., John H. Schwarz, and Edward Witten. *Superstring Theory*. Cambridge, England: Cambridge University Press, 1987.

Gross, David. *Unified Theories of Everything*. Naples: Nella Sede Dell'Istituto, 1989.

Hawking, Stephen W. *A Brief History of Time: From the Big Bang to Black Holes*. New York: Bantam Books, 1988.

Hurn, Jeff. *GPS: A Guide to the Next Utility*. Sunnyvale, Calif.: Trimble Navigation, 1989.

Isham, C. J., R. Penrose, and D. W. Sciama (eds.). *Quantum Gravity 2: A Second Oxford Symposium*. Oxford, England: Clarendon Press, 1981.

Jespersen, James, and Jane Fitz-Randolph. *From Sundials to Atomic Clocks: Understanding Time and Frequency*. New York: Dover, 1982.

Kaku, Michio. *Introduction to Superstrings*. New York: Springer-Verlag, 1988.

Kaku, Michio, and Jennifer Trainer. *Beyond Einstein: The Cosmic Quest for the Theory of the Universe*. New York: Bantam Books, 1987.

Kaufman, William J. *Universe*. New York: W. H. Freeman, 1985.

Kursunoglu, Behram N., and Eugene P. Wigner (eds.). *Reminiscences about a Great Physicist: Paul Adrien Maurice Dirac*. Cambridge, England: Cambridge University Press, 1987.

Landes, David S. *Revolution in Time*. Cambridge, Mass.: The Belknap Press of Harvard University Press, 1983.

Laustsen, Sven, Claus Madsen, and Richard M. West. *Exploring the Southern Sky: A Pictorial Atlas from the European Southern Observatory (ESO)*. New York: Springer-Verlag, 1987.

Lederman, Leon M., and David N. Schramm. *From Quarks to the Cosmos: Tools of Discovery*. New York: Scientific American Library, 1989.

Longair, M. S. *Theoretical Concepts in Physics: An Alter-native View of Theoretical Reasoning in Physics for Final-Year Undergraduates*. Cambridge, England: Cambridge University Press, 1984.

Lorentz, H. A., et al. *The Principle of Relativity: A Collection of Original Memoirs on the Special and General Theory of Relativity*. Translated by W. Perrett and G. B. Jeffery. London: Methuen, 1923.

MacDonald, D. K. C. *Faraday, Maxwell, and Kelvin*. New York: Doubleday, 1964.

Magill, Frank N. (ed.). *The Great Scientists*. Danbury, Conn.: Grolier Educational, 1989.

Marshall, Roy K. *Sundials*. New York: MacMillan, 1963.

Mauldin, John H. *Particles in Nature: The Chronological Discovery of the New Physics*. Blue Ridge Summit, Pa.: TAB Books, 1986.

Millikan, Robert A., et al. *Time and Its Mysteries*. New York: New York University Press, 1936.

Mook, Delo E., and Thomas Vargish. *Inside Relativity*. Princeton: Princeton University Press, 1987.

Morris, Richard. *Time's Arrows: Scientific Attitudes toward Time*. New York: Simon and Schuster, 1985.

Newton, David E. *Particle Accelerators: From the Cyclotron to the Super-Conducting Super Collider*. New York: Franklin Watts, 1989.

Parker, Barry:
Einstein's Dream: The Search for a Unified Theory of the Universe. New York: Plenum Press, 1986.
Search for a Supertheory: From Atoms to Superstrings. New York: Plenum Press, 1987.

Peat, F. David. *Superstrings and the Search for the Theory of Everything*. Chicago: Contemporary Books, 1988.

Schwarz, John H. (ed.). *Superstrings: The First 15 Years of Superstring Theory* (Vol. 1). Philadelphia: World Scientific, 1985.

Seeger, Raymond J. *Men of Physics: Galileo Galilei, His Life and His Works*. London: Pergamon Press, 1966.

Segrè, Emilio. *Enrico Fermi: Physicist*. Chicago: The University of Chicago Press, 1970.

Shallis, Michael. *On Time: An Investigation into Scientific Knowledge and Human Experience*. New York: Schocken Books, 1983.

Smith, Alan (ed.). *The International Dictionary of Clocks*. New York: Exeter Books, 1984.

Snow, C. P. *The Physicists*. Boston: Little, Brown, 1981.

Sobel, Michael I. *Light*. Chicago: The University of Chicago Press, 1987.

Taylor, Edwin F., and John Archibald Wheeler. *Spacetime Physics*. San Francisco: W. H. Freeman, 1963.

Tolstoy, Ivan. *James Clerk Maxwell: A Biography*. Chicago: The University of Chicago Press, 1981.

Trefil, James S. *From Atoms to Quarks: An Introduction to the Strange World of Particle Physics*. New York: Charles Scribner's Sons, 1980.

Turner, A. J. *The Time Museum: Time Measuring Instruments*. Rockford, Ill.: The Time Museum, 1984.

Vanier, Jacques, and Claude Audoin. *The Quantum Physics of Atomic Frequency Standards* (Vol. 2). Bristol, England: Adam Hilger, 1989.

Wheeler, John Archibald. *The Journey into Gravity and Spacetime*. New York: Scientific American Library, 1990.

Whitrow, G. J.:
The Natural Philosophy of Time (2d ed.). Oxford, Eng-

land: Clarendon Press, 1980.

The Nature of Time. New York: Holt, Rinehart and Winston, 1972.

Wilczek, Frank, and Betsy Devine. *Longing for the Harmonies: Themes and Variations from Modern Physics.* New York: W. W. Norton, 1988.

Will, Clifford M. *Was Einstein Right? Putting General Relativity to the Test.* New York: Basic Books, 1986.

Wolf, Fred Alan. *Parallel Universes: The Search for Other Worlds.* New York: Simon and Schuster, 1988.

Periodicals

Adair, Robert K. "A Flaw in a Universal Mirror." *Scientific American,* February 1988.

"Atom Time." *Time,* January 17, 1949.

Bollinger, John J., and David J. Wineland. "Microplasmas." *Scientific American,* January 1990.

Boslough, John.:

"The Enigma of Time." *National Geographic,* March 1990.

"Searching for the Secrets of Gravity." *National Geographic,* May 1989.

"Worlds within the Atom." *National Geographic,* May 1985.

Boyer, Timothy H. "The Classical Vacuum." *Scientific American,* August 1985.

Chanowitz, M. S. "The Z Boson." *Science,* July 6, 1990.

Cole, K. C. "A Theory of Everything." *New York Times Magazine,* October 18, 1987.

Datta, A. "CP-Violation." *2001,* February 1990.

Davies, Paul C. W.:

"Matter-Antimatter." *Sky & Telescope,* March 1990.

"Particle Physics for Everybody." *Sky & Telescope,* December 1987.

DeWitt, Bryce S. "Quantum Gravity." *Scientific American,* December 1983.

Feldman, Gary J., and Jack Steinberger. "The Number of Families of Matter." *Scientific American,* February 1991.

Forman, Paul. "Atomichron®: The Atomic Clock from Concept to Commercial Product." *Proceedings of the IEEE,* July 1985.

Forward, Robert L. "Spinning New Realities." *Science 80,* December 1980.

Freedman, Daniel Z., and Peter van Nieuwenhuizen:

"The Hidden Dimensions of Spacetime." *Scientific American,* March 1985.

"Supergravity and the Unification of the Laws of Physics." *Scientific American,* February 1978.

"A Giant LEP for Mankind." *Economist,* August 19, 1989.

Ginsparg, Paul, and Sheldon Glashow. "Desperately Seeking Superstrings?" *Physics Today,* May 1986.

Glashow, Sheldon. "Quarks with Color and Flavor." *Scientific American,* October 1975.

Goldman, Terry, Richard J. Hughes, and Michael Martin Nieto. "Gravity and Antimatter." *Scientific American,* March 1988.

Green, Michael B.:

"Superstrings." *Scientific American,* September 1986.

"Unification of Forces and Particles in Superstring Theories." *Nature,* April 4, 1985.

Hawking, Stephen W.:

"The Direction of Time." *New Scientist,* July 9, 1987.

"The Edge of Spacetime." *American Scientist,* July-August 1984.

Jackson, J. David, Maury Tigner, and Stanley Wojcicki. "The Superconducting Supercollider." *Scientific American,* March 1986.

Jennings, D. A., et al. "The Continuity of the Meter: The Redefinition of the Meter and the Speed of Visible Light." *Journal of Research of the National Bureau of Standards,* January-February 1987.

Langacker, Paul, and Alfred K. Mann. "The Unification of Electromagnetism with the Weak Force." *Physics Today,* December 1989.

Layzer, David. "The Arrow of Time." *Scientific American,* December 1975.

Lemonick, Michael D. "The Ultimate Quest." *Time,* April 16, 1990.

Lipken, Richard:

"Telling Time." *Insight on the News,* July 9, 1990.

"The Timekeepers." *Insight on the News,* July 9, 1990.

Morrison, Philip. "The Overthrow of Parity." *Scientific American,* April 1957.

Moyer, Albert E. "Michelson in 1887." *Physics Today,* May 1987.

Myers, Stephen, and Emilio Picasso. "The LEP Collider." *Scientific American,* July 1990.

Odenwald, Sten:

"The Planck Era." *Astronomy,* March 1984.

"To the Big Bang and Beyond." *Astronomy,* May 1987.

Penrose, Roger. "Twisting Round Space-Time." *New Scientist,* May 31, 1979.

Peterson, Ivars. "A Different Dimension." *Science News,* May 27, 1989.

Quigg, Chris. "Elementary Particles and Forces." *Scientific American,* April 1985.

Rees, John R. "The Stanford Linear Collider." *Scientific American,* October 1989.

Rothman, Tony. "The Seven Arrows of Time." *Discover,* February 1987.

Schwarz, John H.:

"Completing Einstein." *Science 85,* November 1985.

"Resuscitating Superstring Theory." *The Scientist,* November 16, 1987.

"Superstrings." *Physics Today,* November 1987.

Schwarzschild, Bertram M.:

"Anomaly Cancellation Launches Superstring Bandwagon." *Physics Today,* July 1985.

"CERN Experiment Clarifies Origin of CP Symmetry Violation." *Physics Today,* October 1988.

Schwinger, Julian. "A Path to Quantum Electrodynamics." *Physics Today,* February 1989.

Taubes, Gary:

"Detecting Next to Nothing." *Science 85,* May 1985.

"Everything's Now Tied To Strings." *Discover,* November 1986.

Thomsen, Dietrick E. "Kaluza-Klein: The Koenigsberg Connection." *Science News,* July 7, 1984.

Trefil, James S. "Beyond the Quark." *The New York Times Magazine,* April 30, 1989.

Treiman, S. B. "The Weak Interactions." *Scientific American,* March 1959.

Tryon, Edward P. "Is the Universe a Vacuum Fluctuation?" *Nature,* December 14, 1973.

Vessot, R. F. C., and M. W. Levine. "A Test of the Equivalence Principle Using a Space-Borne Clock." *General Relativity and Gravitation,* February 1979.

Wells, David, and Alfred Kleusberg. "GPS: A Multipurpose System." *GPS World,* January-February 1990.

Wigner, Eugene P. "Violations of Symmetry in Physics." *Scientific American,* December 1965.

Wilczek, Frank. "The Cosmic Asymmetry between Matter and Antimatter." *Scientific American,* December 1980.

Wineland, David J., and Wayne M. Itano. "Laser Cooling." *Physics Today,* June 1987.

Other Sources

Ashby, Neil. "A Tutorial on Relativistic Effects in the Global Positioning System." NIST Contract No. 40RANB9B8112, Department of Physics, University of Colorado, Boulder, February 1990.

Goldman, David T., and R. J. Bell (eds.). *The International System of Units (SI).* Washington, D.C.: U.S. Government Printing Office, 1986.

Thompson, Steven D. *Everyman's Guide to Satellite Navigation.* Washington, D.C.: ARINC Research, 1985.

INDEX

Numerals in italics indicate an illustration of the subject mentioned.

A

Acceleration: Galileo's study, 98, *99;* and gravity, equivalence of, 115-116

Accelerators, particle. *See* Particle accelerators

Alvarez-Gaumé, Luis, 68

Anderson, Carl, 29-30

Antiparticles, 22, 29, 38, *39;* antileptons vs. leptons, 46, *47;* antimuons, *28, 43;* antineutrinos, 23, *28, 39, 40-41, 87;* antiproton formation, *39;* antiproton-proton collisions, *28, 29, 49;* antiquarks, *27, 29, 30, 39,* 46, *47, 49;* anti–X Higgs bosons, *39, 47,* 48; collisions with, in particle accelerators, *28-31;* excess of matter over, 38, 46, 48, 133; in kaon decay, 44, *45;* negative energy states and, 55-56; in weak-force interactions, *43*

Aristotle (Greek philosopher), 10, 93-94

Arrows of time, 123, 125, *126-133;* entropy, 123, *128-129;* expanding vs. contracting universe, 122-123, *130-131;* kaon decay, 44, *45,* 132, *133;* vs. time symmetry, 125, *126-127, 132-133*

Astronomical clock, *95*

Atom: models of, 19-20

Atomic clocks, *108-113;* cesium-beam, *110-111,* 112, 121; first, *97;* hydrogen-maser, *108-109,* 110, 121, *122;* ion-trap, *112-113;* navigation satellites with, *117-120,* 121; relativistic effects measured by, 108, 116, 121, 122

Augustine of Hippo, Saint, 93, 94-95

B

Bare mass, 31

Baryons, types of, 26, *27*

Beta decay, 23, 29; and weak force, 23, 25, 42

Big Bang era, 92; and expanding universe, 122, 130; and GUT, 36-37; Higgs bosons, *32, 39, 46-47,* 48, 133; Inflation Era, 48; matter-antimatter annihilations, 48, *49;* matter-antimatter imbalance, 38, 46, 48, 133; as quantum fluctuation, 56; in supersymmetry theory, 65

Blackbody radiation, 109

Bohr, Niels, 19-20; quoted, 22

Bosons (force carriers), 23, *27, 39;* exchange of, in point-particle view, *87;* gluons, *27, 29,* 34, *39, 49, 87;* Higgs, *32, 39, 46-47,* 48, 133; in string theory, 60, 61, 84, *85;* in supersymmetry, 64-65; weak-force carriers, *27, 28,* 30, *32,* 35, *39, 87. See also* Gravitons; Mesons; Photons

Bradley, James, 103-104

Broken symmetry, 34, 35, 36; in matter-antimatter imbalance, 42, 46, 48, 133; of neutral kaon, 44, 133; parity violations, 42-43, 44, 68, 133; supersymmetry, 65

C

Calabi-Yau spaces, 71

Calculus: Newton and, 8, 100

Celestial mechanics: Newton's theory, 8-9, *13;* time symmetry of, *126-127*

CERN (European Center for Nuclear Research): Large Electron-Positron Collider, *30-31;* Super Proton Synchrotron (SPS), *28,* 30

Cesium-beam clocks, *110-111,* 112, 121

Charge of particles, *chart* 26-27; color, of quarks, 34; symmetry, 132-133

Chirality (handedness) of particle spin, *41,* 42, *43,* 68; and compactification, 68; reversal of, for antiparticles, 43

Chronometer, marine, *96*

Clocks, *94-97,* 96-98; gravity's effect on, 116, *117,* 121, 122; inner works, *90-91;* mechanical, earliest, *95,* 97-98; shadow, *94,* 96; time dilation, effect of, 115, *117,* 121; water, *94,* 96-97, 98, 99. *See also* Atomic clocks

Closed superstrings: heterotic, 70-71, *84-85, 86;* Type II theory, 66-68; world sheets of, 70, *80-81, 86-89*

Cloud chambers: collisions in, *24,* 29

Compactification of dimensions, 65, 67, 68, 70, 71

Cooling laser, *112,* 113

Correlation, principle of, 107

Cosmic rays, *24,* 29

Cyclical view of time, 94, 95

D

Dimensionality, 60; compactification, 65, 67, 68, 70, 71; eight-dimensional space, 64, 73, 76; fifth dimension, 54; limited perspective, problem of, *60-61;* in string theory, 61-62, 67, 68, 70, 71, 84, 85; twenty-six dimensions, positing of, 60

Dirac, Paul: Dirac's equation, 22, 55-56

Drum clock, *95*

Dual symmetry, 43, *132-133*

E

$E_8 x E_8$ symmetry group, 70

Eclipses of Jupiter's moons, 101-102

Eddington, Sir Arthur, 103, 123

Egyptian clocks, *94,* 96

Einstein, Albert, 37, *114;* quoted, 15, 115-116; uncertainty, opposition to, 22, 124. *See also* Relativity

Electricity, 11; as fluid, 17; and magnetism, 10-12, 17-18

Electromagnetic waves, 18; constant speed of, 19

Electromagnetism: development of theory, 10-12, 17-18; and gravity, mathematical unification of, 54; point-particle vs. superstring view of, *87;* quantum electrodynamics, 22-23, 31-32, 55-56; symmetry of,

33, *63;* and weak force, unification of, 34-35, 36. *See also* Photons

Electrons, 18, 19, 20, *26;* in beta decay, 23, 42; collisions with positrons, production of, *30-31;* energy transitions, 20-21, 22; global vs. local transformation, *63;* gravitons and, *16;* from K-long decay, *45;* left- vs. right-handed, *40-43;* from muon decay, *40-41, 42, 87;* and photons, 20, 22, 87; self-energy of, and infinity problem, 23, 31-32; in uncertainty principle, 21, 55; from W− decay, *28*

Electroweak force, 35; in GUT, 36-37; theorists of, 34-35, *36*

Elementary particles. *See* Subatomic particles

Energy: electron transitions, 20-21, 22; and mass, 19; negative states of, 55-56

Entropy, 123, *128-129*

Equivalence, principle of, 116

Ether, 105-106, *107*

European Center for Nuclear Research (CERN): Large Electron-Positron Collider, *30-31;* Super Proton Synchrotron (SPS), *28, 30*

Expanding universe, 122, *130-131;* reversal of, possible, 122-123, 131

F

Faraday, Michael, *11,* 12, 17

Fermi, Enrico, *22,* 23, 25

Fermi National Accelerator Laboratory (FermiLab), near Chicago, Ill.: Tevatron accelerator, *6-7, 29*

Fermions, *26-27;* in string theory, 60, 61; in superstring theory, 84; in supersymmetry, 64-65. *See also* individual names

Feynman, Richard P., 31; quoted, 76-77, 78

Field theory: Maxwell's, 17-18, 19. *See also* Quantum theory

Flatland (two-dimensional space), *60-61*

Flavors of quarks, *27,* 33-34

Force carriers. *See* Bosons

Force lines and force fields, 17, 18

Forces, fundamental, 10, 92; point-particle vs. superstring views of, *87. See also* Electromagnetism; Gravity; Strong force; Symmetry; Weak force

G

Galaxy cluster, *50-51*

Galileo Galilei, 98, *99,* 101

Gamma Draconis (star): shift in position of, 103-104

Gauge invariance. *See* Symmetry

Gell-Mann, Murray, 33, 34, *35,* 61, 69

General theory of relativity, 14, 52, 53-54; acceleration and gravity, equivalence of, 115-116; gravity's effect on clocks, 116, *117,* 121, 122; vs. quantum theory, 10, 16, 53, *72,* 73, 79, 80, 88; shape of universe in, 122; space-time distortion, *14-16,* 52, *88;* vs. twistor theory, *72-75*

Geometry, 56; topology vs., *57,* 58

Georgi, Howard: theory by, 36-37

Ghosts (negative probabilities), 59-60

Glashow, Sheldon, *36;* electroweak theory, 34-35, 36; GUT, 36-37; superstring theory, derogation of, 77-78

Global Positioning System (GPS), 120-121

Global symmetry, *63*

Gluons, *27, 29,* 34, *39, 49,* 87

Grand unification theories (GUTs), 36-37

Gravitation, universal, theory of, 8-9, *13*

Gravitino, 65

Gravitons, *16, 27, 39, 87;* in string theory, 62, *85, 87,* 89; in supersymmetry, 65

Gravity: carriers of (gravitons), *16, 27, 39,* 62, 65, *85, 87,* 89; Newton's theory, 8-9, *13;* string theory and, 62, 67, 89; supergravity, 65; unification with quantum theory, problem of, 10, 37, 53, 54, 72, 79. *See also* General theory of relativity

Gravity wells, *14-16,* 52, *88*

Green, Michael, *66,* 67, *68-70*

Gross, David, 70, *71*

GUTs (grand unification theories), 36-37

H

Hadrons, 33; kinds of, *27;* in string theory, 60, 61; Veneziano's model, 59. *See also* individual names

Handedness. *See* Chirality

Harvey, Jeffrey, 70, *71*

Hawking, Stephen, 123

Heisenberg, Werner, 20-*21,* 23, 55

Heisenberg's uncertainty principle, 21-22, 55, 56, 123-124; and space-time, *16, 72,* 73, *75,* 88

Heterotic strings, 70-71, *84-85;* pants diagram, *86*

Higgs, Peter, 35, 46

Higgs bosons, *32, 39, 46-47,* 48, 133

Hourglasses, *95,* 96

Hydrogen-maser clocks, *108-109,* 110; for testing relativity, 121, *122*

I

Inclined plane: Galileo's experiment with, 98, *99*

Infinities, problem of, 23, 79; and renormalization, 31-32, 35; superstrings as solution, 67, 80

Inflation Era, 48

Infrared lasers in cesium-beam clock, *110, 111*

Interferometer: Michelson-Morley experiment, 105-106, *107*

Intermediate vector bosons, *39,* 87; W and Z, *27, 28, 30, 32,* 35

Inverse-square law, 9, 11, 13

Ion-trap, *112-113*

J

Jennings, Don, *103*

Jupiter (planet): moons, 101-*102*

K

K^0 and $\bar{K}^0$; K_1 and K_2 (kaons), 44, 45

Kaluza, Theodor, *54;* work of, 54-55

Kaluza-Klein theories, 54-55, 65, 68

Kaons (k-mesons), *27, 39;* decay of, *44, 45,* 132, *133*

Kepler, Johannes, 8, 127

Klein, Oskar, *54;* work of, 54-55

Klemola 44 (galaxy cluster), *50-51*

K-long particle, *44, 45*

L

Laplace, Pierre-Simon, 123-124

Large Electron-Positron Collider, *30-31*

Lasers: in cesium-beam clock, *110, 111;* in ion-trap, *112,* 113; in light-speed measurement, 103

Least-action principle: paths following, *80-81*

Lee, Tsung Dao, 34, 42, 43

Leibniz, Gottfried Wilhelm, *100-*101

Leptons, *26, 39;* vs. antileptons, 46, *47;* decay reactions, 23, *28, 40-41, 42, 45. See also* Electrons

Levine, Martin, 121, *122*

Light cones, *72-73, 75*

Light-speed, 92; Bradley's measurement of, 103-104; lantern experiments, 101, 104; in Maxwell's calculations, 18; Michelson-Morley experiment, 105-*107;* recent determinations of, 103, 104-105; Römer's approximation of, 101-102; and special theory of relativity, 19, 114-115

Linear accelerator, *31*

Local symmetry, *63-*64; in supersymmetry, question of, 65

Lorentz, Hendrik Antoon, 107

M

Magnetism, 11; and electricity, 10-12, 17-18

Marine chronometer, *96*

Martinec, Emil, 70, *71*

Masers, hydrogen: atomic clocks, *108-109*, 110, 121, *122*

Mass: bare, 31; density of, in cosmos, 130-131; energy and, 19; of particle types, *chart 26-27*; space-time distortion by, *14-16*, 52, *88*; vibrational states and, 82-83

Matter vs. antimatter. *See* Antiparticles

Maxwell, James Clerk, *11*, 17-18, 19

Mercury (planet): gravity well, *14*; perihelion, precession of, *13, 15*

Mesons, 27, *28-29*, *30-31*, *39*; kaons, *27, 39*, 44, *45*, 132, 133; pions, *24, 27, 39*, 44, *45*; tau and theta, 42

Michelson, Albert, 104, 105-*106*

Michelson-Morley experiment, 105-*107*

Microwaves, atomic clocks using: cesium-beam, *110-111*; hydrogen-maser, *108-109*

Mills, Robert: work of, 32-33

Minkowski, Hermann, *114*, 115

Moon: orbit of, 9, *13*

Moons of Jupiter, 101-*102*

Morley, Edward, *106*; experiment by Michelson and, 105-*107*

Motion and time, analysis of: acceleration, 98, *99*, 115-116; early, 93-94; Newton's, 99-100; time dilation, 114-115, *117*, 121. *See also* Light-speed

Motion of planets. *See* Planetary motion

Multidimensionality. *See* Dimensionality

Muons (mu-mesons), *26*, *30-31*; decay of, *40-41*, 42, *87*; and graviton, *87*; from Z⁰ decay, *28*

N

Nambu, Yoichiro, *58*, 60, 61

Navigation satellites, *117-120*, 121

Negative energy states, 55-56

Negative probabilities (ghosts), 59-60

Neutral kaons, *39*; decay of, 44, *45*, 132, *133*

Neutrinos and antineutrinos, 23, *26*, *39*; from muon decay, *40-41*, 87

Neutrons and protons, 23, 25, *27*

Neveu, André, 61

Newton, Isaac, 8, *9, 100*; and Leibniz, 100-101; time, approach to, 99-100, 101; universal gravitation, theory of, 8-9, *13*

Nuclear forces. *See* Strong force; Weak force

Null lines, *73, 75*

O

Olive, David, 65

Orbiting navigation systems, *117-120*, 121

Orbits: of Mercury, precession of perihelion in, *13, 15*; in space-time, 115; time symmetry of, *126-127*; universal gravitation and, 9, *13*

Ørsted, Hans Christian, 11-12, 18

P

Pants diagram, *86*

Parity, 40-41, 132; violations of, 42-43, 68, 133

Particle accelerators, *28-32*; Large Electron-Positron Collider, *30-31*; Stanford Linear Collider, *31*; Superconducting Supercollider, *32*; Super Proton Synchrotron, *28, 30*; Tevatron, *6-7*, 29

Pauli, Wolfgang, 23; quoted, 77

Pendulums, 98; clocks with, *96, 97*

Penrose, Roger, *76*, 123; twistor theory, 72, *73*, 74, *75, 76*, 77

Perihelion of Mercury: precession, *13, 15*

Petersen, Russell, *103*

Photons, *27, 39*; and electrons, 20, 22, *87*; energy-to-frequency ratio, 53; heterotic strings and, *85*; high-energy, *49*; in hydrogen maser, *108-109*; paths of, *72-73, 75*; scattering of, *111, 112, 113*; virtual, 22-23

Pions (pi-mesons), *24, 27, 39*; kaon decay into, 44, *45*

Planck, Max: work of, 18-19, 20, 53

Planck's constant, 53

Planetary motion: precession of Mercury's perihelion, *13, 15*; time symmetry of, *126-127*; universal gravitation, theory of, 8-9, *13*

Point particles vs. strings, 70, *80-81, 87*

Positrons, 22, 29; decay into, *43*, 44, *45*; electron collisions with, production of, *30-31*

Pound, Robert V.: experiment, 116, 121

Precession of Mercury's perihelion, *13, 15*

"Princeton String Quartet," 70, *71*

Principia (Newton), 9

Proton-antiproton collisions, *49*; in particle accelerators, *28, 29*

Proton decay, predicted, 37

Proton formation from quarks, *27, 39, 49*

Proton-proton collisions, *32*, 50

Protons and neutrons, 23, 25, *27*

Q

Quantum chromodynamics (QCD), 34, 62

Quantum electrodynamics (QED), 22-23, 55-56; and renormalization, 31-32

Quantum fluctuations, virtual particles from, 25, *28-29*, 56, 124; virtual photons, 22-23

Quantum leaps, 20

Quantum of action, 53

Quantum theory (quantum mechanics), 20; comprehensive (Theory of Everything), 53, 79; electroweak, 34-35, 36; fifth dimension, 54; grand unification theory (GUT), 36-37; Heisenberg's theory, 20-21; Planck's work and, 18-19, 20, 53; vs. relativity, 10, 16, 53, *72, 73*, 79, 80, *88*; space, view of, 53, 56; vs. superstring theory, particle interactions in, 70-71, 86, *87*; uncertainty in, *16*, 21-22, 55, 56, *72, 73, 75*, 88, 123-124. *See also* Electromagnetism; Strong force; Subatomic particles; Symmetry; Weak force

Quarks, 26, *27*, 33-34, *39, 87*; in accelerator collisions, *29, 30*; doublets and triplets, kinds of, *27*; gravitons and, *16*; quantum chromodynamics, 34, 62; and string theory, 60-61, *85*; X Higgs decay into, 46, *47, 49*

Quartz-crystal clocks, *97*, 108; cesium-beam, *110-111*; hydrogen-maser, *108-109*

R

Rabi, I. I.: quoted, 43

Ramond, Pierre, 61

Ramond-Neveu-Schwarz theory, 61, 64

Rebka, Glen A.: experiment, 116, 121

Relativity, 52; clocks for measuring, 108, 116, 121, 122; and Dirac's equation, 22; light-speed and, 19, 114-115; and navigation satellites, *117*; and spacefarer scenarios, 116. *See also* General theory of relativity

Renormalization, 31-32, 35

Rohm, Ryan, 70, *71*

Römer, Ole, 101-103, *102*

Rutherford, Ernest, 19-20

S

Salam, Abdus, 35, *36*, 64

Sandglasses, *95*, 96

Satellite navigation systems, *117-120*, 121

Scattering matrix (S-matrix), 59; with twenty-six dimensions, 60

Scattering of photons, *111, 112, 113*

Scherk, Joel, 62, 65

Schrödinger, Erwin, 21, 55

Schwarz, John, 61, 62, *66*, 67, 68-70; quoted, 62, 67, 69, 76

Schwinger, Julian S., 31

141

Second law of thermodynamics: increasing entropy, 123, *128-129*

Self-energy, electron's, and infinity problem, 23; renormalization, 31-32

S-matrix (scattering matrix), 59; with twenty-six dimensions, 60

SO(32) symmetry group, 69

Space, 52; geometry vs. topology in study of, 56, *57*, 58; at quantum level, 53, 56. *See also* Dimensionality; Motion and time

Space shuttle: plotting location of, *118-119*

Space-time, 14, 52; closed strings in, 70-71, *80-81, 86-89*; distortion of, *14-16*, 52, *88*; light cones in, *72-73, 75*; Minkowski concept of, 114, 115; vs. twistor space, *73, 75*

Special theory of relativity, 19, 52, 114-115; and Dirac's equation, 22; time dilation, 114-115, *117*, 121

Spin, particle, *chart* 26-27, 61; chirality (handedness), *41*, 42, 43, 68; heterotic string and, 84

Standing waves (currents) of heterotic superstrings, *84*, 85

Stanford Linear Collider (SLC), 30-*31*

Starlight: shifts in, 103-104, 122

Stellar parallax, 103

Strathdee, John, 64

String theory, 60-62, 80; and gravity, 62, 67, 89; heterotic, 70-71, *84-85, 86*; lack of support for, 62, 65-66, 67, 76-78; originator of (Nambu), *58*, 60, 61; Ramond-Neveu-Schwarz theory, 61, 64. *See also* Superstrings

Strong force, 25, 28, *87*; carriers of, *27*, 28-*29*, 30-31, 34, *39, 49, 87*; gauge field theory, 33; grand unification theory, 36-37; particles subject to (hadrons), 33, 59, 60, 61; quantum chromodynamics, 34, 62

Subatomic particles, 10, *26-27, 39, 87*; in cloud chambers, *24*, 29; in Sun's gravity well, *16*; in supersymmetry, 64-65, 78; from twistors, 74; from vibrations, 58, 60, *82-83. See also* Antiparticles; Bosons; Particle accelerators; Quantum theory; String theory; *individual names*

Sun: gravity well of, *14-16*; Mercury's orbit around, *13, 15*

Superconducting Supercollider (SSC), near Dallas, Tex., planned, *32*

Supergravity, 65

Superparticles, 65, 78

Super Proton Synchrotron (SPS), 28, *30*

Superspace, 64

Superstrings, 53, 65-67, 68-70, 79, *80-89;* anomaly-free symmetry groups, 69, 70; closed, 66-67, 70-71, *80-81, 84-89;* heterotic, 70-71, *84-85, 86;* "Princeton String Quartet," 70, *71;* skepticism about, 76-78; Type I and Type II, 66-67, 68-69; vibrational states of, *82-83;* Witten as supporter of, 67-68, 70, 71, 76, 78; world sheets of, 70, *80-81, 86-89*

Supersymmetry, 63, 64-65; and string theory, 65-67, 80; superparticles, 65, 78

Symmetry, 32-33, 40; broken, 34, 35, 65, 42-43, 44, 46, 48, 68, 133; dual, 43, *132-133;* global, *63;* local, *63-64*, 65; parity, *40-41*, 132; parity violations, 42-43, 68, 133; supersymmetry, 63, 64-65, 78; supersymmetry-string connection, 65-67, 80; time symmetry, 125, *126-127*, 132, 133

Symmetry groups: $E_8 x E_8$, 70; (SO)32, 69; in supersymmetry models, 65; in Type I theory variations, 68-69

T

Tau and theta mesons, 42

Tevatron accelerator, Fermi National Accelerator Laboratory, *6-7*, 29

Theory of Everything (TOE), 53, 79

Thermodynamics, second law of: increasing entropy, 123, *128-129*

Thomson, Sir Joseph John, 18, 19

't Hooft, Gerard, 35

Thought experiments: Einstein's, 19, 114, 116; Schrödinger's, 55; in time reversal, 125, *126-133*

Time, 92-95; absolute zero of, 92, 122; Aristotle's view of, 93-94; Augustine's view of, 93, 94-95; dilation of, 114-115, *117*, 121; Galileo's studies, 98, *99*, 101; Leibniz's view of, 100, 101; Newton's approach to, 99-100, 101; reversal of, vs. arrows of time, 122-123, 125, *126-133. See also* Clocks; Light-speed; Space-time

Time dilation, 114-115, *117*, 121

Time symmetry, 125, *126-127*, 132, 133

Timing laser, *112*, 113

TOE (Theory of Everything), 53, 79

Tomonaga, Shin'ichirō, 31

Top-down reasoning, 59, 60

Topology, *57*, 58; of strings, 70-71, *86-89*

Transformations, Lorentz, 107

Triangulation, *118*

Tryon, Edward P., 56

Twistor theory, 72, *73*, 74, *75, 76*, 77

Type I and Type II superstring theories, 66-67, 68-69

U

Uncertainty principle, 21-22, 55, 56, 123-124; and space-time, *16, 72, 73, 75*, 88

Universal gravitation, theory of, 8-9, *13*

Universe: closed vs. open-ended, 122; contraction of, possible, 122-123, 131; expanding, 122, *130-131. See also* Big Bang era

V

Veneziano, Gabriele, 59; mathematical model by, 59-60, 70

Vessot, Robert, 121, *122*

Vibrations of strings, 58, 60, *82-83*

Virtual particles, 25, 28-29, 56, 124; photons, 22-23

W

W+ and W− bosons, *27*, 28, *32*

Wallingford astronomical clock, *95*

Water clocks, *94*, 96-97; Galileo's use of, 98, 99

Wave equations for electrons, 21

Weak force, 23, 25, 39; antimuon decay, *43;* carriers of, *27*, 28, 30, *32*, 35, *39, 87;* dual symmetry of, 43; and electromagnetism, unification of, 34-35, 36; kaon decay, 44, *45*, 132, *133;* muon decay, *40-41*, 42, *87;* parity violation by, 42-43, 68

Weinberg, Steven, 35, *36*, 92, 122

Wess, Julius, *64*

Wheeler, John Archibald, 58, 124

Witten, Edward, *67-68*, 70, 71, 76, 78

World lines: world points and, 115; vs. world sheets, 70, *80-81*

World sheets, 70, *80-81, 87, 88-89;* pants diagram, *86*

Wormholes, 58

Wrist watch, *97*

X

X Higgs boson, *39, 46-47*, 48, 133

Y

Yang, Chen Ning: symmetry, work on, 32-33, 34, 42, 43

Yau, Shing-Tung, 71

Yukawa, Hideki, *23*, 25; mesons predicted by, 28-29, 30, 31

Z

Z^0 bosons, *27, 28, 30*

Zero mode (vibrational state), 83

Zumino, Bruno, *64*

Zweig, George, 33, *35*, 61

ACKNOWLEDGMENTS

The editors wish to thank Karon L. Anderson, The Time Museum, Rockford, Ill.; Michael S. Chanowitz, Lawrence Berkeley Laboratory, Berkeley, Calif.; Bryce DeWitt, University of Texas, Austin; Robert Drullinger, National Institute of Standards and Technology, Boulder, Colo.; Val Fitch, Princeton University, Princeton, N.J.; Terry Goldman, Los Alamos National Laboratories, Los Alamos, N.M.; David Gross, Princeton University, Princeton, N.J.; Rolf Krauss, Ägyptisches Museum, Berlin, Germany; Dorothy A. Mastricola, The Time Museum, Rockford, Ill.; Jacques-Clair Noens, Observatoire du Pic-du-Midi, France; John H. Schwarz, California Institute of Technology, Pasadena, Calif.; Lee Smollen, Syracuse University, Syracuse; Robert Switzer, Mathematisches Institut, Universität Göttingen, Germany; Marie-Josée Vin, Observatoire de Haute Provence, France; Fred L. Walls, National Institute of Standards and Technology, Boulder, Colo.; Julius Wess, Sektion Physik, Universität Munchen, Munich, Germany; Richard M. West, European Southern Observatory, Munich, Germany; David J. Wineland, National Institute of Standards and Technology, Boulder, Colo.; Bryant Winn, Aerospace Corporation, El Segundo, Calif.; Bruce Winstein, University of Chicago, Ill.

PICTURE CREDITS

The sources for the illustrations that appear in this book are listed below. Credits from left to right are separated by semicolons, from top to bottom by dashes:

Cover: Lawrence Berkeley Laboratory/Science Photo Library, London. Front and back endpapers: Art by Time-Life Books. 6, 7: David Parker/SPL/Photo Researchers. 8: Initial cap, detail from pages 6, 7. 9: Reproduced by the Gracious Permission of Her Majesty the Queen. 11: Smithsonian Institution, Pic. No. 78-1831; The Master and Fellows of Peterhouse, Cambridge. 13-16: Art by Yvonne Gensurowsky. 21: The Bettmann Archive. 22, 23: Enrico Persico, courtesy Rosa Segre; Segre Collection, AIP Niels Bohr Library. 24: C. Powell, P. Fowler, and D. Perkins/SPL/Photo Researchers. 26, 27: Art by Al Kamajian. 28-32: Art by Karen Barnes of Stansbury, Ronsaville, Wood, Inc. 35: Los Alamos National Laboratory, courtesy George Zweig; California Institute of Technology. 36: Herman J. Kokojan/Black Star; © Nubar Alexanian; International Centre for Theoretical Physics, Rome. 39-49: Art by Al Kamajian. 50: European Southern Observatory, Garching, Germany, inset by Fermi National Accelerator Laboratory. 52: Initial cap, detail from pages 50, 51. 54: Mathematisches Institut, Universität Göttingen, Germany; AIP Niels Bohr Library. 57: Art by Fred Holz. 58: The University of Chicago. 61: Art by Matt McMullen. 63: Art by Fred Holz. 64: University of California, Lawrence Berkeley Laboratory; Vidyut Jain, Max Planck Institut für Physik und Astrophysik, Munich. 66, 67: Rene Sheret; Judah Passow/ J. B. Pictures; *The Star Ledger*, courtesy The Institute for Advanced Study, School of Natural Sciences, Princeton University. 71: Boston University Photo Services; Mark Steinmetz, University of Chicago; Michael Pirrocco, Princeton University; Robert P. Matthews, Princeton University. 72-75: Art by Stephen R. Bauer. 76: Anthony Howarth, Woodfin Camp & Associates. 80-89: Art by Matt McMullen. 90, 91: Heinz Zinram, courtesy Society of Antiquaries, London. 92: Initial cap, detail from pages 90, 91. 94, 95: Background art by Yvonne Gensurowsky. Ägyptisches Museum, Staatliche Museen Preussischer Kulturbesitz, Berlin, Foto Margarette Büsing; courtesy of The Time Museum, Rockford, Illinois (2); Science Museum, London; Heinz Zinram, courtesy Society of Antiquaries, London. 96, 97: Background art by Yvonne Gensurowsky. Courtesy of The Time Museum, Rockford, Illinois (2); Heinz Zinram, courtesy National Maritime Museum, Greenwich; courtesy of The Time Museum, Rockford, Illinois; National Institute of Standards and Technology, Gaithersburg, Md.; Seiko Corporation, Tokyo. 99: Computer art by Time-Life Books; Scala/Art Resource, New York. 100: National Portrait Gallery, London; Granger Collection, New York. 102, 103: Ole Rømers Museum, Tåstrup, Denmark; National Maritime Museum, London; National Institute of Standards and Technology. 106, 107: Clark University Archives; Case Western Reserve University, courtesy AIP Niels Bohr Library; Special Collections Department, Nimitz Library, U.S. Naval Academy, courtesy Hale Observatories. 108-113: Art by Stephen R. Wagner. 114: Courtesy AIP Niels Bohr Library; Suddeutscher Verlag Bilderdienst, Munich. 117-120: Art by Rob Wood of Stansbury, Ronsaville, Wood, Inc. 122: Smithsonian Astrophysical Observatory. 126-133: Art by Stephen R. Wagner.

Time-Life Books is a division of Time Life Inc., a wholly owned subsidiary of
THE TIME INC. BOOK COMPANY

TIME-LIFE BOOKS

PRESIDENT: Mary N. Davis

Managing Editor: Thomas H. Flaherty
Director of Editorial Resources:
Elise D. Ritter-Clough
Director of Photography and Research:
John Conrad Weiser
Editorial Board: Dale M. Brown, Roberta Conlan, Laura Foreman, Lee Hassig, Jim Hicks, Blaine Marshall, Rita Thievon Mullin, Henry Woodhead
Assistant Director of Editorial Resources/ Training Manager: Norma E. Shaw

PUBLISHER: Robert H. Smith

Associate Publisher: Trevor Lunn
Editorial Director: Donia Steele
Marketing Director: Regina Hall
Production Manager: Marlene Zack
Supervisor of Quality Control: James King

Editorial Operations
Production: Celia Beattie
Library: Louise D. Forstall
Computer Composition: Deborah G. Tait (Manager), Monika D. Thayer, Janet Barnes Syring, Lillian Daniels
Interactive Media Specialist: Patti H. Cass

Correspondents: Elisabeth Kraemer-Singh (Bonn), Christine Hinze (London), Christina Lieberman (New York), Maria Vincenza Aloisi (Paris), Ann Natanson (Rome). Valuable assistance was also provided by Elizabeth Brown, Katheryn White (New York), Judy Aspinall (London).

VOYAGE THROUGH THE UNIVERSE

SERIES EDITOR: Roberta Conlan
Series Administrator: Norma E. Shaw

Editorial Staff for *Workings of the Universe*
Art Directors: Barbara M. Sheppard (principal), Dale Pollekoff
Picture Editor: Kristin Baker Hanneman
Text Editors: Robert M. S. Somerville (principal), Stephen Hyslop
Assistant Editors/Research: Dan Kulpinski, Patricia A. Mitchell
Writers: Mark Galan, Darcie Conner Johnston
Editorial Assistant: Katie Mahaffey
Copy Coordinator: Juli Duncan
Picture Coordinator: David Beard

Special Contributors: J. Kelly Beatty, Mark Bello, Robert Cane, George Constable, James Dawson, Marge duMond, Eliot Marshall, Chuck Smith, Mark Washburn (text); Vilasini Balakrishnan, Mary Mayberry, Dara Norman, Eugenia Scharf (research); Barbara L. Klein (index).

CONSULTANTS

ROBERT KEMP ADAIR is a Sterling professor of physics at Yale University in New Haven, Connecticut. His special fields of interest are elementary particle physics and nuclear physics.

DAVID ALLAN is a physicist at the Time and Frequency Division of the National Institute of Standards and Technology in Boulder, Colorado. His work focuses on atomic time scale, and the classification of the statistical characteristics of atomic and molecular frequency standards.

NEIL ASHBY, a professor of physics at the University of Colorado in Boulder, Colorado, studies theoretical physics. He researches transport phenomena, quantum statistical mechanics, and relativity.

S. JAMES GATES, JR. is a professor of physics at the University of Maryland in College Park, Maryland, and Howard University in Washington, D.C. His work includes quantum field theory and supersymmetrical theories.

GARY T. HOROWITZ, a professor of physics at the University of California at Santa Barbara, specializes in developing a better understanding of the phenomenon of gravity.

EMIL MOTTOLA is a physicist at the Theoretical Division of the Los Alamos National Laboratory in Los Alamos, New Mexico. His studies include matter and antimatter as well as quantum gravity and the cosmological constant.

STEN ODENWALD is an infrared astronomer with the Space Sciences Division of the Naval Research Laboratory in Washington, D.C. He also teaches at the Smithsonian and has published widely on cosmology.

DON NELSON PAGE is a professor of physics at the University of Alberta in Edmonton. He researches black holes and the early universe.

ROGER SHERMAN is a science historian and curator at the Department of Electricity and Modern Physics at the National Museum of American History in Washington, D.C., where he cares for the Modern Physics collection. He has also been involved with the museum's exhibition "Atomic Clocks."

FRANK WILCZEK is a professor of physics at the Institute for Advanced Study in Princeton, New Jersey. His research includes high energy physics and quantum field theory.

Library of Congress Cataloging in Publication Data
Workings of the universe / by the editors of Time-Life Books.
p. cm. (Voyage through the universe).
Bibliography: p.
Includes index.
ISBN 0-8094-6916-2.
ISBN 0-8094-6917-0 (lib. bdg.).
1. Space and time. 2. Matter.
3. Quantum theory. 4. Astrophysics.
I. Time-Life Books. II. Series.
QC173.59.S65W67   1990
530.1'1—dc20   90-47596 CIP

For information on and a full description of any of the Time-Life Books series, please call 1-800-621-7026 or write:
Reader Information
Time-Life Customer Service
P.O. Box C-32068
Richmond, Virginia 23261-2068

Earth: diameter 7,926 miles

Neptune: diameter 30,775 miles

Uranus: diameter 31,763 miles

Red supergiant: diameter 400 million miles

Solar System: diameter 7.5 billion miles

Globular cluster: diameter 2×10^{14} miles

Milky Way: diameter 100,000 light-years

Local Group of galaxies: 6 million light-years across

Largest double radio source: length 17 million light-years